Multiorganizational Arrangements for Watershed Protection

With cross-pollination of the public administration and policy implementation literatures, Madeleine Wright McNamara and John Charles Morris present the Multiorganizational Interaction Model as a framework to explore the use of cooperation, coordination, and collaboration between 15 federal/state agencies, local governments, and nongovernmental organizations working together to restore coastal habitats and replenish aquatic resources on Virginia's Eastern Shore.

Content analysis of data collected through interviews and organizational documents allows comparisons to be made regarding the distribution of data across the continuum of interaction. The presence of policy mandates intending to prescribe relationships, coupled with strong perceptions of collaboration, create opportunity to explore mandated and voluntary collaboration. Themes regarding mapping relationships within the multiorganizational arrangement, movement on the continuum, and implementation through mid-level personnel are discussed. The combination of theory development and testing provides readers with a theoretical framework through which to think about interorganizational interactions, and a case study to illustrate the ways in which these complex relationships manifest themselves in practice.

Multiorganizational Arrangements for Watershed Protection will be essential for scholars, students, and policy makers.

Madeleine Wright McNamara is an adjunct assistant professor in the School of Public Service at Old Dominion University. She served previously as a Visiting Assistant Professor in the Department of Political Science at the University of New Orleans and as the Waterways Management Coordinator for the U.S. Coast Guard's Eighth District in New Orleans. Her research interests include collaboration, public policy, and interorganizational theory. Her work appears in journals such as *Public Works Management &*

Policy, the *International Journal of Public Administration*, *Policy & Politics*, and the *Journal for Nonprofit Management*, among others. In addition, she authored chapters in *Speaking Green with a Southern Accent: Environmental Management and Innovation in the South* (2010) and *Advancing Collaboration Theory: Models, Typologies, and Evidence* (2016; Routledge).

John Charles Morris is a professor in the Department of Political Science at Auburn University. He has studied collaboration, public-private partnerships, and state comparative policy for more than twenty-five years, and has published widely in public administration and public policy. Dr. Morris has a significant number of publications. He is the co-editor of *Speaking Green with a Southern Accent: Environmental Management and Innovation in the South* (2010), and *True Green: Executive Effectiveness in the US Environmental Protection Agency* (2012). Dr. Morris is also the co-editor of *Building the Local Economy: Cases in Economic Development*, published by the Carl Vinson Institute of Government, University of Georgia, in 2008, and is the co-editor of a three-volume series (2012) on prison privatization, titled *Prison Privatization: The Many Facets of a Controversial Industry*. His most recent co-authored books include *The Case for Grassroots Collaboration: Social Capital and Ecosystem Restoration at the Local Level* (2013); *Advancing Collaboration Theory: Models, Typologies, and Evidence* (2016; Routledge, co-edited with Katrina Miller-Stevens); *State Implementation of the Affordable Care Act: Choices and Decisions* (2019, Routledge); and *Organizational Motivation for Collaboration: Theory and Evidence* (2019, with Luisa Diaz-Kope). In addition, he has published more than sixty-five articles in refereed journals, and nearly forty book chapters, reports, and other publications.

Routledge Research in Public Administration and Public Policy

Multiorganizational Arrangements for Watershed Protection

Working Better Together

**Madeleine Wright McNamara
and John Charles Morris**

Routledge
Taylor & Francis Group

NEW YORK AND LONDON

First published 2021
by Routledge
52 Vanderbilt Avenue, New York, NY 10017

and by Routledge
2 Park Square, Milton Park, Abingdon, Oxon, OX14 4RN

Routledge is an imprint of the Taylor & Francis Group, an informa business

Library of Congress Cataloging-in-Publication Data
A catalog record for this title has been requested

ISBN: 978-0-367-48641-9 (hbk)
ISBN: 978-1-003-04205-1 (ebk)

Typeset in Times New Roman
by codeMantra

This book is dedicated to
Cecilia, Rosaline, Theodore, Silas, and Oliver

Contents

Figures

Tables

Abbreviations

CPT   Coastal Policy Team
DEQ   Virginia Department of Environmental Quality
GEMS   Geospatial and Educational Mapping System
IAM   Interorganizational Arrangement Model
MIM   Multiorganizational Interaction Model
NOAA   National Oceanic and Atmospheric Administration
PDC   Planning District Commission
VCZM   Virginia Coastal Zone Management

Acknowledgments

As with any large research project, this book is only possible thanks to the kindness, openness, and contributions of many people. To begin with, we wish to extend sincere thanks to the members of Virginia's Coastal Zone Management Program, and their government and nongovernmental partners, who gave so generously of their time to help us understand the importance and complexity of natural resource preservation. Their stories and experiences contributed greatly to this work and enhanced significantly our understanding of the interactions that sustain the ecosystem on Virginia's Eastern Shore. We are also grateful for the generosity and enthusiasm of Natalja Mortensen and Charlie Baker for their insightful direction and support throughout this process. Finally, we thank our families for their love and support.

1 Multiorganizational Interactions for Watershed Protection

Organizations often face implementation challenges compounded by inherently complex and interconnected policy problems (O'Toole, 2000). Increased demands on government, coupled with fewer resources, further exacerbate the inabilities of individual organizations to independently implement public policy. As a result, public organizations work across bureaucratic boundaries to increase government's capacity for addressing complex problems (Kettl, 2003; Mandell, 1999). Partnerships between public, private, and nonprofit organizations develop and interdependencies form. Multiorganizational arrangements are increasingly used as agencies work together to implement policy by diversifying resources and expertise (Hall & O'Toole, 2004; Keast, Mandell, Brown, & Woolcock, 2004; O'Toole, 1993). Implementation inevitably requires interactions across organizational boundaries (Hjern & Porter, 1981; O'Toole, 1993). Therefore, it is important to expand our understanding of the interactions that take place between organizations when working together to implement policy. This study introduces the Multiorganizational Interaction Model (MIM) to explore the use of cooperation, coordination, and collaboration during multiorganizational policy implementation.

The policy implementation literature identifies two main theoretical approaches to policy implementation: the top-down approach and the bottom-up approach. These two approaches address implementation in different ways and emphasize different values (Schofield, 2001). Much of the implementation research focuses on identifying variables specific to each approach (see, for example, Mazmanian & Sabatier, 1989; O'Toole, 1986). The circuitous debate comparing the strengths and weaknesses of the top-down and bottom-up approaches must be replaced with research identifying the variables most critical to policy implementation (O'Toole, 1986, 2000). The number of variables, wide variation in their perceived importance, and complexity of interactions are problematic for theoretical advancement (Goggin, 1986; O'Toole, 1986, 2000); conceptual clarity remains elusive. Although theorists acknowledge the importance of reconciling both approaches, a synthesized framework based on the combined strengths of the top-down and bottom-up approaches is needed (O'Toole, 2000; Saetren, 2005).

Organizations are often mandated by legislation to implement public policy in support of particular policy goals. Some mandates require two or more organizations to work together during policy implementation (see, for example, Caruson & MacManus, 2006; O'Toole & Montjoy, 1984; Raelin, 1982). In these instances, relationships are likely to develop between organizations. The importance of interaction across organizational boundaries is first acknowledged in Pressman and Wildavsky's (1973) *Implementation*, where ignorance of organizational interdependence in complex decision chains ultimately contributes to a mismatch between policy expectations and outcomes. Despite this recognition more than three decades ago, little is done to examine the interactions during multiorganizational implementation empirically. The focus of this research is on the different types of interactions that occur when organizations work together to implement public policy.

A Critique of Traditional Approaches to Multiorganizational Implementation

The policy literature recognizes the importance and complexity of multiorganizational implementation (see, for example, O'Toole, 1986; Pressman & Wildavsky, 1973). However, empirical inquiry emphasizes formalized interactions based on policy mandate, agency rulemaking, or organizational procedures (Caruson & MacManus, 2006; Raelin, 1982). More specifically, the literature focuses on the extent to which policies identify interorganizational partners (Hall & O'Toole, 2004), policy characteristics that induce or constrain interdependence (May, 1995; O'Toole, 1995; O'Toole & Montjoy, 1984), or the structures used in multiorganizational implementation (Mandell, 1994; O'Toole, 1997). Despite a nonhierarchical nature, much of the literature involving implementation networks also emphasizes formalized interactions based on organized efforts (Hall & O'Toole, 2004; Mandell, 1994). The common thread among these different approaches to the literature is that organizations are assumed to work together in a formalized arrangement based on a requirement to do so.

There are three problems with this approach. First, it fails to consider the possibility that multiorganizational implementation may occur outside the boundaries of operational authority. Informal interactions between organizations are important (Keast et al., 2004) and should be examined empirically. Second, legislators are limited in their abilities to foresee and specify the interactions required in complex implementation settings (O'Toole & Montjoy, 1984). It is highly unlikely that policy mandates account for all potential interactions within the implementation settings. Third, the literature appears not to have progressed beyond the top-down/bottom-up debate even though O'Toole (2000) declared its ending nearly a decade ago. Exclusive acknowledgment of formalized arrangements advocates a top-down approach. In the multiorganizational implementation literature, this approach is emphasized under the guise of alternative nomenclature—coordination.

Collective action lies at the heart of multiorganizational implementation (O'Toole, 1991). Researchers use terms such as cooperation, coordination, or collaboration to describe interactions in multiorganizational implementation (Jennings & Ewalt, 1998; Lundin, 2007; O'Toole & Montjoy, 1984). While these terms are often used within the public administration, organization theory, and education literatures (see, for example, Intriligator, 1992; Keast, Brown, & Mandell, 2007; Mattessich, Murray-Close, & Monsey, 2001), there is much that still needs to be done to understand empirically the nuances that distinguish these terms from one another.

A lack of conceptual clarity impacts inquiry in three ways. First, there is a tendency to broadly categorize interaction terms with little regard for the definitions that distinguish them from one another (Caruson & MacManus, 2006; Keast, Brown, & Mandell, 2007; Mandell & Steelman, 2003). Without acknowledging and defining each term, theorists cannot properly consider the range of interactions potentially useful in multiorganizational implementation settings. Second, the appropriate application of cooperation, coordination, and collaboration cannot be acknowledged when they are used interchangeably. Misapplication of terms makes it difficult to identify the conditions under which it is appropriate to use a particular type of interaction as an implementation strategy. Third, researchers often assume an interaction occurs even though its presence is not empirically tested (see, for example, Bryson, Crosby, & Stone, 2006; Imperial, 2001; Jennings & Krane, 1994; Kettl, 2003). The collective impact from these problems further perpetuates theoretical inconsistency. A conceptual model pertaining to the use of different types of interactions during multiorganizational implementation is nonexistent.

The purpose of this book is to explore interactions between organizations when working together to implement policy. More specifically, this research explores the use of cooperation, coordination, and collaboration between government and nongovernmental organizations during implementation of the Virginia Seaside Heritage Program. This study of multiorganizational implementation is important for two reasons. First, a major contribution of this study is the introduction of the Multiorganizational Interaction Model (MIM). The strength of the MIM is that it resolves an earlier model's ambiguities by clearly distinguishing between operationalizations of cooperation, coordination, and collaboration based on application of the policy implementation and interorganizational theory literatures. These distinctions provide opportunities for consistent theoretical inquiry and improve the model's transferability for future use in alternative settings.

Second, this is the first time that a model focused on multiorganizational interactions is applied to a public policy implementation setting. Broadening the scope of current inquiry to explore different types of interactions may improve our theoretical understanding of policy implementation in multiorganizational arrangements. A continuum of interactions gives researchers a different way of looking at the top-down/bottom-up debate while moving beyond a narrow reconciliation of the two approaches. This application to public policy expands the use of interorganizational theory and suggests that both literatures may benefit from collective inquiry. For example, empirical research on informal interactions fills a gap in the current policy implementation and interorganizational theory literatures, both of which emphasize formal interactions deliberately configured to attain policy or organizational goals.

With this in mind, this study examines three research questions:

1   Does the MIM help explain interactions in a policy implementation setting?
2   How do administrators perceive the use of cooperation, coordination, or collaboration when working in a multiorganizational arrangement to implement policy?
3   How are multiorganizational interactions initiated?

The first two research questions explore the applicability of the MIM and the perceived use of different types of interactions in a multiorganizational implementation setting. Variables within all four constructs of the model are explored to address these two research questions. The third research question explores whether multiorganizational interactions are initiated formally through legislative mandate or agency rulemaking, informally through street-level experience or common interests, or a combination of both. The impetus for collective action variable within the interorganizational policy objective construct and the formality of the agreement variable within the interorganizational infrastructure construct of the MIM are explored to address this research question.

A Model of Multiorganizational Interaction

The MIM provides the basis for the conceptual framework within this study. A previous version of this model, developed in the health education literature, examines multiorganizational arrangements in settings such as health and human service delivery, medical and social service provision, and education (see, for example, Edmondson, 2006; LaRocco, 1997; Thatcher, 2007). The design, assumptions, construct nomenclature, and operationalizations from previous versions of the model are transformed in this study to eliminate ambiguities and align with the policy implementation and interorganizational theory literatures. This transformation results in the development of the MIM.

The MIM is a theoretical lens that can be used to explore a continuum of interactions between organizations. Cooperation, coordination, and collaboration are the interaction terms used to describe this continuum. These terms, and their placement along a continuum, are recognized by some researchers in the public administration, organization theory, and education literatures (see, for example, Intriligator, 1992; Keast, Brown, & Mandell, 2007; Mattessich, Murray-Close, & Monsey, 2001).

At one end of the spectrum, cooperation represents an interaction between independent organizations who can individually accomplish the task at hand but voluntarily and informally work together to build capacity or serve individual interests in pursuit of simple goals (Keast, Brown, & Mandell, 2007; Mattessich, Murray-Close, & Monsey, 2001; O'Leary & Bingham, 2007). At the other end of the spectrum, collaboration represents an interaction between organizations with collective responsibility for interconnected tasks who work together voluntarily or by mandate in pursuit of complex goals which cannot be accomplished by a single organization and are based on shared interests (Gray, 1989; Keast, Brown, & Mandell, 2007; Mattessich, Murray-Close, & Monsey, 2001; Thomson & Perry, 2006). Coordination is placed in the middle of the two end points and represents an interaction between organizations requiring some assistance from other organizations to accomplish individual missions in which formal linkages are mobilized voluntarily or by mandate in pursuit of multifaceted goals that support common objectives (Keast, Brown, & Mandell, 2007; Mattessich, Murray-Close, & Monsey, 2001).

Contribution of the Research

Broadening the scope of current inquiry to explore different types of interactions may improve our understanding of multiorganizational policy implementation. Linkages with interorganizational theory can be established to guide the implementation literature toward a fourth generation of research. Distinguishing between different types of interactions and identifying the conditions that warrant their use may further our understanding of how particular interactions can be used as implementation strategies to help multiorganizational arrangements fulfill policy goals. An approach that considers a continuum of interactions as implementation strategies gives researchers a different way of looking at the top-down/bottom-up debate while moving beyond a narrow reconciliation of the two approaches. More specifically, a continuum of interactions bridges the two approaches by allowing researchers to look at the nature of the interaction without having to assume that one approach is more important to implementation than the other. The formulation of previous implementation models relies considerably on the preconceived notions of their developers and their chosen approach to implementation (O'Toole, 1993; Schofield, 2001).

Since the top-down and bottom-up approaches are often seen to embrace competing variables, the selection of one approach over another involves a value judgment based on a limited set of variables considered to be most relevant. These assumptions limit the researcher to a narrow view of implementation by allowing them to ignore the relative importance of variables associated with the other approach (O'Toole, 1993). On the other hand, the MIM does not make any presumptions regarding variable importance, and each approach is considered equally important in analyzing the interactions between organizations during implementation. In moving beyond the constraints inherent to the top-down and bottom-up approaches, researchers are not restricted to a particular set of variables and may be able to see implementation for what it really is.

Interactions are formed in a variety of ways (Robinson, 2006). Legislation and agency rulemaking do not necessarily capture the operational patterns of interaction that actually occur during implementation (Hall & O'Toole, 2004). The MIM can be used to further research informal interactions between organizations. If informal interactions are empirically identified, two assumptions must be reconsidered. The first assumption involves the multiorganizational implementation literature. Theorists can no longer assume that formal, coordinated strategies are the only way in which organizations interact when implementing policy in multiorganizational arrangements. The second assumption involves interorganizational theory. Researchers can no longer assume that all interactions are deliberately and formally configured by planning personnel to align with the type of interaction specified in the interorganizational objective (see, for example, Thatcher, 2007). The potential for informal interactions to occur in multiorganizational arrangements must be considered.

The Virginia Coastal Zone Management Program

The Virginia Coastal Zone Management (VCZM) Program provides the setting for this study. Its mission is to protect, restore, and strengthen Virginia's coastal areas by managing and overseeing activities that affect coastal resources. The VCZM Program is a network of Virginia state agencies and local governments who administer state laws, regulations, and policies to protect coastal resources. This network of organizations is selected as the setting for this study based on their involvement in implementing a policy mandate, a need for them to work collectively to implement this mandate, and a potential for a variety of interactions to occur between the government and nongovernmental organizations involved in implementing the program.

The VCZM Program was established in 1986, by executive order, to protect Virginia's coastal zone and in response to the Coastal Zone Management Act of 1972 (Kaine, 2006; United States Congress, 1972). The executive order explains the mission of the VCZM Program, specifies policy goals, identifies the Virginia Department of Environmental Quality (DEQ) as the lead agency, and requires specific state agencies to participate in program implementation.

Despite its designation as lead agency, the DEQ does not have control over other state agencies.

State agency involvement in the VCZM Program is specified by executive order and classified into two categories: (1) agencies primarily responsible for implementing the VCZM Program's enforceable policies and (2) agencies responsible for assisting with the VCZM Program (Kaine, 2006). These agencies are listed in Table 1.1.

In Virginia's eight coastal areas, local governments are involved in the VCZM Program through the Planning District Commissions (PDCs). The PDCs focus on coastal management issues of greater than local concern by facilitating relationships, passing information, and pooling resources between state and local governments (Office of Ocean and Coastal Resource Management [OCRM], 2004). On an annual basis, the VCZM Program awards each PDC a grant to provide local governments with technical assistance (OCRM, 2004; Virginia Coastal Zone Management Program [VCZMP], 2008a).

State agencies and PDCs within the VCZM Program partner with nongovernmental organizations, such as nonprofits and private businesses, to implement coastal programs or policies (see, for example, VCZMP, 2006). Partnerships with nonprofit organizations include The Nature Conservancy and Eastern Shorekeeper. Partnerships with private organizations include Cherrystone Aquafarms and Southeast Expeditions. Nonprofit organizations and private businesses are not identified in the executive order. Typically, nongovernmental organizations do not receive direct grant funding through the VCZM Program.

Table 1.1 Virginia State Agencies Designated by Executive Order

Agencies Primarily Responsible for Implementation of Enforceable Policies	Agencies Responsible for Assisting with the Program
Department of Environmental Quality	Department of Historic Resources
Department of Conservation and Recreation	Department of Forestry
Marine Resource Commission	Department of Agriculture and Consumer Services
Department of Game and Inland Fisheries	Virginia Institute of Marine Science
Department of Health	Department of Transportation
	Virginia Economic Development Partnership

The Coastal Policy Team

The VCZM Program established the Coastal Policy Team (CPT) to provide a forum to bring organizations together to develop and implement coastal policies, discuss coastal resource issues, and resolve conflicts (OCRM, 2004). It is comprised of representatives from key state agencies and each of the eight coastal PDCs. Each member of the CPT has voting rights; decisions, such as prioritizing issues and funding strategies, are based on consensus. The CPT provides policy recommendations to the VCZM Program staff (VCZMP, 2005).

Members of the CPT have access to the coastal geospatial and educational mapping system (GEMS). This web-based planning tool helps them share information and align decisions pertaining to land use and resource management. The VCZM Program contributes considerable funding to this system in order to facilitate communication and informed decision making among partner organizations. The accuracy of the system relies on the data quality provided by the members of the CPT. In addition to potentially aligning local implementation efforts, another goal of GEMS is to better inform policy decision making by strengthening linkages between local land use plans and the state's water use policies (OCRM, 2007). Efforts to train representatives from key agencies and PDCs on GEMS are currently underway.

Grant Funding and the Virginia Coastal Zone Management Program

The VCZM Program staff administers grant money funded by the National Oceanic and Atmospheric Administration (NOAA). These funds are used to maintain ongoing programs, support a large program identified as a main focal area, or help smaller projects get started. The VCZM Program found it beneficial to fund a long-term project aligned with their main focal area, which is selected every three years (OCRM, 2004). When identifying long-term projects, input is solicited from organizations within the network and other nongovernmental partners (VCZMP, 2005).

Grant contracts are used to distribute money, define the scope of a particular project, and formally identify a single organization's responsibility for meeting the requirements in the grant. Funding is available to coastal states with federally approved coastal management programs. A lead organization is designated for each project and becomes legally responsible for implementing the specifications within the grant contract. This organization has discretion to work with other government agencies and nongovernmental partners to achieve project goals; these relationships are not specified in the grant and may occur more informally. Partnering organizations are often involved in project implementation even if they do not receive grant funding from the VCZM Program. Each project is assigned a grant coordinator and a project manager from the staff of the VCZM Program. The grant coordinator ensures the grant money is used as intended. The project manager may act as a facilitator between the program and the lead organization responsible for local project implementation. Project management typically goes beyond the terms specified in the grant contract. While grant funding is important, it is one part of the overall strategy for a particular initiative.

The Virginia Seaside Heritage Program

In 2002, the Virginia Seaside Heritage Program became the main focal area for the VCZM Program (VCZMP, 2007). As a result, this program receives significant funding and coastal management expertise from the VCZM Program; support is scheduled to continue through September 2008. The Virginia Seaside Heritage Program, and more specifically the interactions that occur between the organizations involved in implementing this program, provide the focus for this study.

The primary goals of the Virginia Seaside Heritage Program are to restore coastal habitats and replenish aquatic resources along Virginia's Eastern Shore while promoting sustainable economic activities such as ecotourism and aquaculture (VCZMP, 2007, 2008c). Aquatic resources include underwater grasses, oysters, scallops, finfish, waterfowl, and shorebirds (OCRM, 2007). The presence of these aquatic resources within the waters surrounding Virginia's Eastern Shore are dramatically declining due to over-harvesting, disease, and habitat loss (VCZMP, 2007, 2008c). In addition to habitat restoration, aquaculture, and ecotourism, another goal of the Virginia Seaside Heritage Program is to draft an agreement between key players promoting management strategies for sustaining coastal resources (VCZMP, 2008b).

A network of federal agencies, Virginia state agencies, local governments, and nongovernmental organizations implement the policies associated with the Virginia Seaside Heritage Program. These organizations are identified in Table 1.2. It is through these

Table 1.2 Network of Organizations in the Virginia Seaside Heritage
Program

Organizational Type	*Specific Organizations in the Network*
Federal Agencies	U.S. Fish and Wildlife Service
Virginia State Agencies/ Programs	Department of Environmental Quality
	Coastal Zone Management Program
	Marine Resources Commission
	Department of Conservation and Recreation
	Department of Game and Inland Fisheries
Local Government	Accomack County
	Northampton County
	Accomack-Northampton PDC
Nongovernmental Organizations	The Nature Conservancy
	Eastern Shorekeeper
	Southeast Expeditions
	Cherrystone Aquafarms
	College of William & Mary[a]
	Institute of Marine Science[a]
	Center for Conservation Biology[a]
	University of Virginia[a]

a Although these academic institutions are state sponsored, they operate autonomously as individual organizations.

partnerships that coastal habitats are restored; aquatic resources are replenished; and economic development is managed.

Layout of the Book

This chapter introduces the problem, provides relevant background information highlighting gaps in the current research, and identifies the purpose of this research. The next chapter presents the Multiorganizational Interaction Model and offers discussion of relevant policy implementation and interorganizational theory literature.

Chapter 3 focuses on analyzing the data collected from interviews and organizational documents while presenting an empirical test of the model developed in the prior chapter. The chapter also offers a brief discussion of the methods used in this study. In Chapter 4, perceptions of administrators implementing coastal resource policies on Virginia's Eastern Shore and the formality of interactions among these administrators are explored as a further triangulation of the model. We conclude the book with a discussion pertaining to the implications of the study for both theory and practice.

References

Bryson, J., Crosby, B., & Stone, M. (2006). The design and implementation of cross-sector collaborations: Propositions from the literature. *Public Administration Review*, 66. 44–55.

Caruson, K., & MacManus, S. (2006). Mandates and management challenges in the trenches: An intergovernmental perspective on Homeland Security. *Public Administration Review*, 66(4). 522–536.

Edmondson, B. (2006). Factors that contributed to the longevity of a coordinated school health program in a northeastern state. *ProQuest Dissertations & Theses Full Text*, 67(03). (UMI No. 3209926)

Goggin, M. (1986). The "too few cases/too many variables" problem in implementation research. *The Western Political Quarterly*, 39(2). 328–347.

Gray, B. (1989). *Collaborating: Finding common ground for multiparty problems*. San Francisco, CA: Jossey-Bass Publishers.

Hall, T., & O'Toole, L. (2004). Shaping formal networks through the regulatory process. *Administration and Society*, 36(2). 186–207.

Hjern, B., & Porter, D. (1981). Implementation structures: A new unit of administrative analysis. *Organization Studies*, 2(3). 211–227.

Imperial, M. (2001). Collaboration as an implementation strategy: An assessment of six watershed management programs. *Dissertations & Thesis Full Text*, 62(02), 767. (UMI No. 3005481)

Intriligator, B. A. (1992). *Establishing inter-organizational structures that facilitate successful school partnerships*. Paper presented at the Annual Meeting of the American Education Research Association Boston, MA, April 16–20, 1990. (ERIC Document Reproduction Service No. ED 347 692).

Jennings, E., & Ewalt, J. (1998). Interorganizational coordination, administrative consolidation, and policy performance. *Public Administration Review*, 58(5). 417–428.

Jennings, E., & Krane, D. (1994). Coordination and welfare reform: The quest for the philosopher's stone. *Public Administration Review*, 54(4). 341–348.

Kaine, T. (2006). *Executive order number twenty-one*. Retrieved February 4, 2008, from http://www.deq.state.va.us/coastal/exorder.html

Keast, R., Brown, K., & Mandell, M. (2007). Getting the right mix: Unpacking integration meanings and strategies. *International Public Management Journal*, 10(1). 9–33.

Keast, R., Mandell, M., Brown, K., & Woolcock, G. (2004). Network structures: Working differently and changing expectations. *Public Administration Review*, 64(3). 363–371.

Kettl, D. (2003). Contingent coordination: Practical and theoretical puzzles for homeland security. *American Review of Public Administration*, 33(3). 253–277.

LaRocco, D. (1997). An analysis of the collaborative nature of state and local Part H interagency efforts and the consequent relationships between these levels of Part H governance. *Dissertations & Thesis Full Text*, 58(04), 1245. (UMI No. 9728849)

Lundin, M. (2007). Explaining cooperation: How resource interdependence, goal congruence, and trust affect joint actions in policy implementation. *Journal of Public Administration Research and Theory*, 17(4). 651–672.

Mandell, M. (1994). Managing interdependencies through program structures: A revised paradigm. *American Review of Public Administration*, 24(1). 99–121.

Mandell, M. (1999). The impact of collaborative efforts: Changing the face of public policy through networks and network structures. *Policy Studies Review*, 16(1). 4–17.

Mandell, M., & Steelman, T. (2003). Understanding what can be accomplished through interorganizational innovations: The importance of typologies, context, and management strategies. *Public Management Review*, 5(2). 197–224.

Mattessich, P., Murray-Close, M., & Monsey, B. (2001). *Collaboration: What makes it work?* Saint Paul, MN: Amherst H. Wilder Foundation.

Mazmanian, D., & Sabatier, P. (1989). *Implementation and Public Policy.* New York: University of California.

Office of Ocean and Coastal Resource Management. (2004). *Evaluation findings for the Virginia Coastal Management Program: November 1999 through July 2003.* Washington, DC: National Oceanic and Atmospheric Administration.

Office of Ocean and Coastal Resource Management. (2007). *Final evaluation findings for the Virginia Coastal Zone Management Program: August 2003 through May 2006.* Washington, DC: National Oceanic and Atmospheric Administration.

O'Leary, R., & Bingham, L. (2007). Conclusion: Conflict and collaboration in networks. *International Public Management Journal*, 10(1). 103–109.

O'Toole, L. (1986). Policy recommendations for multi-actor implementation: An assessment of the field. *Journal of Public Policy*, 6(2). 181–210.

O'Toole, L. (1991). Multiorganizational policy implementation: Some limitations and possibilities for rational choice contributions. *Paper for Workshop on Games in Hierarchies and Networks,* Koln, Germany.

O'Toole, L. (1993). Interorganizational policy studies: Lessons drawn from implementation research. *Journal of Public Administration Research and Theory*, 3(2). 232–251.

O'Toole, L. (1995). Rational choice and policy implementation: Implications for interorganizational network management. *American Review of Public Administration*, 25(1). 43–57.

O'Toole, L. (1997). Treating networks seriously: Practical and research-based agendas in public administration. *Public Administration Review*, 57(1). 45–52.

O'Toole, L. (2000). Research on policy implementation: Assessment and prospects. *Journal of Public Administration Research and Theory*, 10(2). 263–288.

O'Toole, L., & Montjoy, R. (1984). Interorganizational policy implementation: A theoretical perspective. *Public Administration Review*, 44(6). 491–503.

Pressman, J., & Wildavsky, A. (1973). *Implementation* (3rd ed.). Berkeley: University of California Press.

Raelin, J. A. (1982). A policy output model of interorganizational relations. *Organization Studies*, 3(3). 2243–267.

Robinson, S. (2006). A decade of treating networks seriously. *The Policy Studies Journal*, 34(4). 589–598.

Saetren, H. (2005). Facts and myths about research on public policy implementation: Out-of-fashion, allegedly dead, but still very much alive and relevant. *Policy Studies Journal*, 33(4). 559.

Schofield, J. (2001). Time for a revival? Public policy implementation: A review of the literature and an agenda for future research. *International Journal of Management Reviews,* 3(3). 245–263.

Thatcher, C. (2007). A study of interorganizational arrangement among three regional campuses of a large land-grant university. *Dissertations & Thesis Full Text*, 68(03). (UMI No. 3255178)

Thomson, A., & Perry, J. (2006). Collaboration processes: Inside the black box. *Public Administration Review*, 55. 20–32.

United States Congress. (1972). *Coastal Zone Management Act of 1972.* Retrieved February 15, 2008 from http://coastalmanagement.noaa.gov/czm/czm_act.html

Virginia Coastal Zone Management Program. (2005). *Final draft: Section 309 needs assessment.* Richmond, VA: Author.

Virginia Coastal Zone Management Program. (2006). *Memorandum of understanding: Relating to the management of conservation lands located on the southern tip of the Eastern Shore.* Richmond, VA: Author.

Virginia Coastal Zone Management Program. (2007). *Virginia Seaside Heritage Program: Goals and project highlights 2002 through 2007.* Richmond, VA: Author.

Virginia Coastal Zone Management Program. (2008a). *PDC technical assistance grant minimum standards.* Richmond, VA: Author.

Virginia Coastal Zone Management Program. (2008b). *Virginia's Eastern Shore seaside management plan draft.* Richmond, VA: Author.

Virginia Coastal Zone Management Program. (2008c). *Virginia Seaside Heritage Program: Accomplishments 2002 through 2008.* Richmond, VA: Author.

2 The Multiorganizational Interaction Model

Implementation research has spawned three generations, with much emphasis placed on reconciling the top-down and bottom-up approaches (Elmore, 1985; Goggin, Bowman, Lester, & O'Toole, 1990; Matland, 1990, 1995). Inherent differences between these two approaches make reconciliation difficult, and circuitous discussions largely prevent identifying the variables most critical to policy implementation (Menzel, 1987; O'Toole, 1986, 2000). As a result, conceptual clarity is elusive. An emphasis on multiorganizational implementation may lead to a refocus of the literature and move it toward a fourth generation of research (Imperial, 2001).

Multiorganizational implementation becomes more important as public organizations face increasing demands, fewer resources, and complex policy problems (O'Toole, 1993, 1997). Working together to meet policy and program goals is well documented as a valuable governance strategy (see, for example, Ansel & Gash, 2008; Bryson, Crosby, & Stone, 2015; Emerson, Nabatchi, & Balogh, 2012; among others). In addition, the benefit of dialogue in achieving mutually agreed upon goals to resolve small aspects of larger problems seems like a realistic strategy in today's interconnected environment (Hardy, Lawrence, & Grant, 2005; Morris, Gibson, Leavitt, & Jones, 2013). Despite its importance, empirical inquiry pertaining to multiorganizational implementation primarily assumes organizations work together in mandated, formalized, or hierarchical arrangements (see, for example, Caruson & MacManus, 2006; Hall & O'Toole, 2000, 2004; Raelin, 1982). For example, a mandate may be used to develop formal authority to support policy objectives and prescribe processes for a hierarchically tiered arrangement within a multiorganizational governance structure (Brummel, Nelson, & Jakes, 2012). However, today's public administrators must find ways to harness interaction across hierarchical operating systems. Arrangements that are not mandated, less formal, or nonhierarchical are not considered empirically within the policy implementation literature despite their utilization in other disciplines. This is problematic as the interorganizational theory literature explores varying relationships that harness interaction across hierarchical operating systems. "Professionals' collaboration" (Nylen, 2007), "contingent coordination" (Kettl, 2003), and "mandated collaboration" (McNamara, 2016a) are some interactions that combine elements of nonhierarchical and hierarchical frameworks.

This study links the policy implementation and interorganizational theory literatures to explore interactions during multiorganizational policy implementation. The first part of this chapter focuses on the policy implementation literature while the second part focuses on the interorganizational theory literature. The final section of this chapter presents the Multiorganizational Interaction Model (MIM); this theoretical lens guides inquiry throughout this book.

Policy Implementation

Implementation research may be organized into three generations (see Goggin et al., 1990). The first generation, roughly the period from 1973 to 1978, focuses on highly descriptive studies implementing authoritative mandates regarding a single policy decision (Matland, 1990). First-generation researchers primarily employ a case study design and routinely focus on implementation failures within individual organizations. They often face critique for their pessimistic view of implementation success (Goggin, 1986) and their inabilities to contribute to a more general theory of implementation (deLeon & deLeon, 2002). Despite these shortcomings, it is important to acknowledge that this research generates a variety of lessons learned, highlighting the complexities inherent in implementation. By recognizing difficulties in translating policy into action, researchers became aware of gross inaccuracies surrounding their perceptions that implementation occurs automatically after policy decisions are made. As a result, researchers widely acknowledge the necessity for further study of the implementation process.

The second generation, roughly the period from 1978 to 1985, focuses on comprehensive theoretical models to highlight two approaches to policy implementation: the top-down approach and the bottom-up approach (Matland, 1990). These two approaches address implementation in very different ways (Schofield, 2001). An inherent emphasis on competing values, such as bureaucratic authority versus local discretion, is often referred to as an implementation paradox (Alexander, 1989; Long & Franklin, 2004; Stoker, 1991).

The search for a conceptual framework that synthesizes the top-down and bottom-up approaches continues throughout the third generation of implementation research. However, synthesizing both approaches into an all-encompassing conceptual framework is complicated by the competing ways in which each approach views implementation. In order to appreciate these complications, the specific set of assumptions that guide each approach must be understood.

Top-Down/Bottom-Up Approaches

The top-down approach focuses on the attainment of centralized objectives. Policy designers play a key role in policy implementation and are assumed to have abilities to impose policy (Linder & Peters, 1987; Mazmanian & Sabatier, 1989; Sabatier, 1986). Policy characteristics and hierarchical controls are critically important to top-down research (Mazmanian & Sabatier, 1989; McFarlane, 1989). By focusing on technical criteria such as statutory clarity, rule promulgation, policy specificity, and hierarchical monitoring, theorists assume that implementation problems can be minimized through careful planning (see, for example, Long & Franklin, 2004; Montjoy & O'Toole, 1979; Van Horn, 1979). Centralized control, clear direction, and authoritative monitoring are needed to ensure local actors implement policy congruent with the goals of policy designers (Edwards & Sharkansky, 1978; O'Toole, 1993; Sabatier, 1986). The top-down approach may be critiqued because it does not recognize differentiation of implementation based on street-level discretion (Long & Franklin, 2004; Schofield, 2001). In addition, today's public administrators are likely to engage in relationships required through mandate but lacking the accountability mechanisms that would traditionally be utilized in a single hierarchical arrangement. A top-down approach does not acknowledge properly the dilemmas administrators face in situations where competing values exist (McNamara, 2016a).

On the other hand, the bottom-up approach focuses on policy implementation influenced by policy actors, local initiatives, citizen needs, and contextual factors (see Maynard-Moody & Musheno, 2000). Rather than being centrally controlled, actors throughout the lowest levels of an organization are afforded discretion to implement policy (Carrington, 2005; Linder & Peters, 1987; Lipsky, 1980). Administrators use personal judgment to make decisions at the lowest levels (Carrington, 2005); these decisions cumulatively create public policy (Lipsky, 1980). As a result, policy continuously evolves and is refined by interactions at various levels (Bovens & Zouridis, 2002; Linder & Peters, 1987). The bottom-up approach may be critiqued because it fails to recognize the potential for policy characteristics or hierarchical controls to influence the local policy environment (Schofield, 2001). In addition, this approach does not acknowledge that interactions may not develop organically (McNamara, 2016a).

While the search for a conceptual framework that synthesizes the top-down and bottom-up approaches continues throughout the third generation of implementation research, much of this research focuses on identifying variables specific to each approach (Mazmanian & Sabatier, 1989; O'Toole, 1986). For example, O'Toole (1986) identifies over 300 variables discussed within the literature for their potential to impact policy implementation. The number of variables identified in implementation research is problematic for three reasons. First, much of the implementation research focuses on case studies. When the number of variables significantly overwhelms the number of cases, the data may incorrectly appear to support inferences. Too much information is being used to explain a small number of cases (Goggin, 1986). Second, it is nearly impossible to measure the specific effect each variable has on the implementation process while accounting for numerous interactions. Third, the quantity of variables blinds researchers from seeing what is truly important.

The circuitous debate comparing the strengths and weaknesses of the top-down/bottom-up approaches must be replaced with research identifying the variables most critical to policy implementation. This is made more difficult by a lack of cumulative research (Hjern, 1982; O'Toole, 1986, 2000). The number of variables, wide variation in their perceived importance, and complexity of interactions paralyze theoretical advancement (Goggin, 1986; Goggin et al., 1990; O'Toole, 1986, 2000). Conceptual clarity remains elusive in the absence of a synthesized framework based on the combined strengths of the top-down and bottom-up approaches (Matland, 1995; O'Toole, 1991; Saetren, 2005).

Reconciling Top-Down/Bottom-Up Approaches

Researchers acknowledge the importance of reconciling the top-down and bottom-up implementation approaches (see Elmore, 1985; Goggin et al., 1990; among others). Some of the more common models are based on the contributions of Elmore (1985), Goggin et al. (1990), and Matland (1990, 1995). Elmore (1985) applies forward and backward mapping to implementation research. By looking at both ends of the implementation process, policy analysts can explore which approach leads to a more advantageous solution (Elmore, 1985). Goggin et al. (1990) use a systems approach to explore how implementers at the state level are influenced by inducements and constraints from the top (federal) and bottom (state and local) levels of government. Their "Communications Model" is intended to explore the complexities of intergovernmental implementation by addressing various components at the federal, state, and local levels (Goggin et al., 1990). Matland (1990, 1995) uses a contingency approach to explore adaptive implementation. According to this model, an implementation strategy is dependent on environmental conflict and statutory ambiguity. By dichotomously organizing these variables into a typology, four types of implementation strategies are identified (Matland, 1995).

Their attempts to synthesize the top-down and bottom-up approaches are commendable, but these models do not account for the multiorganizational arrangements frequently used to implement policy (see, for example, Hjern & Porter, 1981; Keast, Mandell, Brown, & Woolcock, 2004; Lundin, 2007). In addition, an emphasis on competing implementation approaches minimizes efforts placed on developing other research areas deserving of inquiry (Schofield, 2001). As governments face increasingly interconnected problems, it is inevitable that implementation will occur in a pluralistic environment requiring interaction across organizational boundaries (Hjern, 1982; Menzel, 1987; O'Toole, 1993; Robinson, 2006). Implementation success or failure will rely heavily on the organizations involved in policy implementation and the interdependencies between these organizations (Alexander, 1989; O'Toole, 1995).

Interorganizational Implementation

O'Toole (1995) defines interorganizational implementation as two or more organizations working together to implement public policy. This subset of implementation literature, also referred to as multiorganizational implementation, conceptually recognizes the importance and complexity of joint action (see, for example, Elmore, 1985; O'Toole, 1991; Pressman & Wildavsky, 1973). However, little emphasis is placed on developing a conceptual model pertaining to the use of different types of multiorganizational interactions during implementation. While interaction terms such as cooperation, co-ordination, and collaboration are clearly defined in the interorganizational theory literature (Gray, 1989; Mattessich, Murray-Close, & Monsey, 2001), they are arbitrarily used in the multiorganizational implementation literature. Furthermore, the nuances that distinguish these terms are ignored (Keast, Brown, & Mandell, 2007). There is a need to understand how these interactions impact policy implementation in a multiorganizational setting.

Empirical inquiry within the interorganizational implementation literature emphasizes formalized interactions based on policy mandate, agency rulemaking, or organizational procedures. More specifically, the literature focuses on the extent to which policies identify interorganizational partners (Hall & O'Toole, 2000, 2004), policy characteristics that induce or constrain interdependence (May, 1995; O'Toole, 1983; O'Toole & Montjoy, 1984), or the structures used in multiorganizational implementation (Hall & O'Toole, 2000, 2004; O'Toole, 1989, 1997; Raelin, 1982). These inquiries may help researchers gain some knowledge pertaining to interactions during multiorganizational implementation, but an emphasis on formalized interactions only addresses one portion of the larger picture in which informal interactions also occur.

Despite the nonhierarchical nature of network arrangements, much of the literature involving implementation networks also emphasizes formalized interactions based on organized efforts (see, for example, Mandell, 1994; Raelin, 1982). This emphasis is ironic when considering the definition of networks. According to Hall and O'Toole (2004), networks are defined as "two or more units in which not all major components are encompassed within a single hierarchical array" (p. 187). Although other than formal relationships seem likely to occur when organizations work outside their hierarchical boundaries, researchers largely ignore these relationships when examining network arrangements. Despite a definition to the contrary, implementation networks are treated by researchers as an extension of the organizational hierarchy abiding by specifications dictated through policy mandates. For example, deLeon and Varda (2009) identify a theme of heterogeneity within network participants but make no mention of a coupling mechanism to bring participants together in a way that best addresses collective goals.

The common thread among these different approaches to empirical inquiry within the interorganizational implementation literature is that organizations are assumed to work together in a formalized arrangement based on a requirement to do so. Unlike the collaboration literature that focuses on engaging stakeholders in relationships based on the perception of a win-win situation of mutual benefit (McNamara, 2016b), the policy literature emphasizes the power of persuasion within the policy subsystem through the agenda-setting process (Howlett, 2003). Influential relationships play an important role but within the context of overlapping interests within the policy subsystem (McNamara, 2016b).

There are three problems with this approach. First, the possibility that multiorganizational implementation may occur outside the boundaries of operational authority is not considered. An emphasis on formal interactions superficially endorses top-down inquiry, which is supported under the guise of alternative nomenclature— coordination. It is assumed that all levels of the organization comply with the type of interaction specified in the mandate, that organizations work together because they are required to do so, and that local dynamics can be ignored. Researchers fail to consider the relationships that informally develop outside of the organizational hierarchy despite their potential impact on multiorganizational implementation. Informal interactions between organizations are important (Keast et al., 2004; McNamara, 2012), and the ways in which different arrangements form should be examined empirically (Robinson, 2006).

Second, an emphasis on formal interactions incorrectly assumes that relationships between organizations can be predetermined, centrally controlled, and monitored to meet policy goals. This approach requires the variables most important to multiorganizational relationships during policy implementation to be theoretically identified and empirically examined. An exclusively top-down approach fails to consider that legislators are limited in their abilities to foresee and specify the interactions required in complex implementation settings (O'Toole & Montjoy, 1984). Once again, informal interactions may play an important role in multiorganizational implementation; it would be difficult to accurately capture them in the policy mandate (McNamara, 2016a). Therefore, it is highly unlikely that policy mandates can account for all potential interactions within implementation settings. The interorganizational theory literature uses the terms "collaborative entrepreneur" (McNamara, 2016b) or "convener" (McNamara & Morris, 2012) to develop a coupling mechanism that can bring participants together in a way that meets collective interests. Such a role is not utilized in the policy implementation literature.

Third, a formal approach to organizational arrangements within the implementation literature perpetuates the top-down/bottom-up debate even though O'Toole (2000) declared an ending to this debate two decades ago. It is one thing to theoretically support one approach over another after careful consideration, but it is entirely different when one approach is supported based on ignorance of the other. Resolution of the top-down/bottom-up debate requires equal consideration of both approaches. Policy implementation involves more than carrying out a combination of statutory clauses (see, for example, Pressman & Wildavsky, 1973); it requires an understanding of the linkages that occur between the organizations responsible for implementation (Hjern, 1982). Elements of both approaches are needed to examine fully the relationships between organizations because they are impacted by top-down characteristics such as mandate characteristics in addition to bottom-up characteristics such as interorganizational dynamics within the local implementation environment. Therefore, examination of multiorganizational interactions should occur without assuming that one approach is more important to implementation than the other.

Collective action lies at the heart of multiorganizational implementation (O'Toole, 1991). Researchers often use terms such as cooperation, coordination, or collaboration to describe interactions in multiorganizational implementation (see, for example, Jennings & Ewalt, 1998; Lundin, 2007; Mandell & Steelman, 2003; May, 1995; O'Toole, 1983; Robinson, 2006). While these terms are more clearly defined in the interorganizational theory and education literatures (see, for example, Intriligator, 1992; Thomson & Perry, 2006), the nuances that distinguish these terms go unnoticed when applied in a multiorganizational implementation setting. As a result, interaction terms are used arbitrarily and interchangeably to describe relationships within the implementation literature. To date, researchers have not linked cooperation, coordination, or collaboration collectively to multiorganizational implementation. The development of a conceptual model that distinguishes between each type of interaction may broaden our understanding of multiorganizational implementation and help identify the appropriate application for each type of interaction within this setting.

Interorganizational Theory

Today's public managers often face complex social problems that do not abide by bureaucratic boundaries (Kettl, 2003). Interdependencies between government agencies and nongovernmental partners can be used to generate a variety of multiorganizational arrangements. These arrangements are also referred to as "interorganizational innovations" and may increase government's capacity for action (Mandell & Steelman, 2003). The terms used most commonly in the public administration literature to describe multiorganizational interactions are cooperation, coordination, and collaboration (see, for example, Agranoff, 2006; Caruson & MacManus, 2006; McNamara, 2012; Thomson & Perry, 2006). It is important for practitioners and scholars to understand each of these terms and how they differ from one another (Mandell & Steelman, 2003). A particular type of interaction is not better than another as the contextual environment should dictate the appropriate type of interaction (McNamara, 2016a). Therefore, the following section broadly describes each of these terms and their placement along a continuum of interaction.

The Continuum of Interaction

Some public administration theorists describe cooperation, coordination, and collaboration as falling along a continuum of increased interaction (Bryson, Crosby, & Stone, 2006; Keast, Brown, & Mandell, 2007; Mattessich, Murray-Close, & Monsey, 2001; McNamara, 2012; Thomson & Perry, 2006). At one end of the spectrum, cooperation is an interaction between independent organizations who can individually accomplish the task at hand but may voluntarily and informally work together within existing organizational structures and policies to build capacity or serve individual interests in pursuit of simple, short-term goals (O'Leary & Bingham, 2007a). The desire to work together may be triggered by changes in external factors and the desire to avoid negative impacts associated with these factors (Ospina & Yaroni, 2003). There is no need to define a mission, structure, or planning effort common to the organizations within the arrangement (Mattessich, Murray-Close, & Monsey, 2001). Cooperation may take place without involving organizational leaders (Keast, Brown, & Mandell, 2007).

While cooperation is identified as an interaction term on the continuum, literature pertaining to cooperation is elusive. The few articles that appear to focus solely on cooperative interactions are plagued by a lack of definition, misapplication of the term, or a tendency to interchange cooperation with other interaction terms such as coordination and collaboration (see, for example, Althaus & Yarwood, 1993; Callahan, 2007; Lambright, 1997). The research conducted by Ospina and Yaroni (2003) on labor management provides one exception to these problems. In studies where cooperation is defined and identified as one of three terms on the continuum of interaction, empirical research is lacking (see, for example, O'Leary & Bingham, 2007a). Efforts to decipher cooperation from other interaction terms are addressed in the research conducted by Keast, Brown, and Mandell (2007) regarding administrators' perceptions of differences between the terms cooperation, coordination, and collaboration. However, research solely focused on cooperative interactions and its application to public organizations has yet to develop fully in the interorganizational theory literature.

Coordination is placed in the middle of the two end points. It is an interaction that links organizations in specific areas because some assistance from other organizations is needed to accomplish the individual mission (Jennings, 1994; Jennings & Ewalt, 1998). Agencies typically use this type of interaction to pursue longer-term goals based on repeatable tasks (Mattessich, Murray-Close, & Monsey, 2001). The desire to work together may be voluntary or mandated based on a benefit to achieving individual and compatible mission areas that support common objectives. Although leaders within the individual organizations retain authority over decision making, there may be some overlap in resources, infrastructure, and procedures. Therefore, coordination often implies the need for some shared planning where roles and responsibilities are formally defined (Mattessich, Murray-Close, & Monsey, 2001).

Coordination is typically characterized by instrumental processes that rely on formally structured relationships and hierarchical control to link the infrastructures of individual organizations (see, for example, Mandell, 1994; Van de Ven, Delbecq, & Koenig, 1976; Van de Ven & Walker, 1984). As a result, coordination is often perceived as a formal approach to interaction based on a requirement for organizations to work together (Keast, Brown, & Mandell, 2007). However, less formal ways to view coordination are acknowledged in the literature (Chisholm, 1989; Kettl, 2003; McNamara, 2016a; Wise, 2006).

Van de Ven and Walker (1984) use the term "mobilization coordination" to describe less formal, ad hoc relationships in their longitudinal study of early childhood development organizations. Based on a questionnaire administered to 14 agency directors, they conclude that the use of formal or informal approaches to coordination largely depends on the types of resources used to create interdependencies between organizations. An interdependence based on financial resources generates a more formal approach to coordination; on the other hand, a more informal approach to coordination occurs when aligning resources pertaining to client referrals (Van de Ven & Walker, 1984). A less formal approach to coordination is also recognized by Kettl (2003); he uses the term "contingent coordination" to describe a flexible and adaptable network approach to coordination. McNamara, Morris, and Mayer (2014) also used this term to explore coordinative relationships with flexibility to respond to emergent conditions. Mandell and Steelman (2003) also differentiate between "intermittent coordination" and "regular coordination" (p. 203). According to Chisholm (1989), an informal approach to coordination may initially be facilitated by more formal organizational policies and activities.

The limitation with much of this subset of the literature is that interactions other than coordination are not often considered. Without this consideration, it is difficult to determine the existence of an informal approach to coordination. Perhaps the informal approach to coordination may be better explained by a different interaction term altogether.

At the other end of the spectrum, collaboration is based on interdependence among multiple organizations that share responsibility for interconnected tasks and work together collectively to pursue complex goals that cannot be accomplished by a single organization (Gray, 1989; McNamara, 2012). This type of interaction typically requires great levels of commitment as stakeholders within a particular problem domain frequently interact to develop shared norms, rules, and processes used to make collective decisions (Wood & Gray, 1991). Collaboration is further characterized by organizations that establish a collective unit in which individual organizations relinquish some autonomy to develop new infrastructure and procedures to support mutually beneficial interactions in which decisions are made jointly (Intriligator, 1992, 1994; Mandell, 1994). As a result, organizational boundaries are often blurred (Keast, Brown, & Mandell, 2007) and significant efforts are needed to align planning efforts (Mattessich, Murray-Close, & Monsey, 2001). This process goes well beyond the instrumental approach emphasized by coordination (McNamara, 2012).

Collaboration is not appropriate for use in all situations (McNamara, Miller-Stevens, & Morris, 2019). Collaborative interactions may be most appropriate under certain conditions: when other types of interaction have failed; when complex situations of crisis occur; when problems are so interconnected that responsibility is shared; or when there is a win-win situation based on mutual interest (Bryson, Crosby, & Stone, 2006; Imperial, 2005; Keast et al., 2004). Organizations are limited in their abilities to exploit their collaborative capacity when faced with different statutory responsibilities, different constituencies, competing interests, a lack of slack resources, procedural rigidity, institutional or budgetary constraints, power asymmetries, ideological differences, power disparities, or a history of conflict (Huxham, 2003; Imperial, 2001; McNamara, Miller-Stevens, & Morris, 2019).

It may be particularly difficult for public organizations to sustain collaborative relationships because they are often faced with conventional bureaucratic systems that do not inherently accommodate shared power and joint decision making (McNamara, 2016a). Keast, Brown, and Mandell (2007) use interviews and focus groups to gather information from 40 practitioners pertaining to their understanding of cooperation, coordination, and collaboration. Their findings suggest a tendency among practitioners to revert to coordinated interactions. A majority of their participants, representing multiple levels of government, indicate that collaborative interactions are beneficial but very difficult to sustain within public organizations (Keast, Brown, & Mandell, 2007).

In order to sustain a collaborative relationship, actors representing each organization within the arrangement must have discretion to negotiate rules and make organizational decisions based on the evolution of group deliberation (Mattessich, Murray-Close, & Monsey, 2001; McNamara, 2012). According to Lipsky (1980), the theory of discretion explains how decisions are made at the lower levels of an organization to cumulatively create public policy. Therefore, discretionary judgment is at odds with the command and control authority inherent to bureaucratic organizations. An emphasis on stovepipe specializations, hierarchical structures, and formal governance mechanisms may make it difficult for administrators to obtain the discretion needed to make decisions within the collaborative arrangement and sustain horizontal relationships.

Empirical Differentiation of Terms

Researchers often ignore the differences between cooperation, coordination, and collaboration because empirical inquiry into these interaction terms is largely undeveloped within the interorganizational theory literature. In fact, only a handful of articles within the literature address empirical differences between these terms (see, Keast et al., 2004; McNamara, 2012, 2016b; McNamara, Morris, & Mayer, 2014). In the research conducted by Keast, Brown, and Mandell (2007), interaction terms are distinguished across five dimensions: time to establish the interaction, goals, structural linkages, formality, and risks or rewards. McNamara (2008) expands this research to develop a more detailed continuum of interaction across 18 dimensions. A detailed conceptual framework would be helpful in understanding empirically the different elements within each interaction term.

A lack of conceptual clarity related to interaction terms impacts the interorganizational theory literature in three ways. First, there is a tendency to broadly categorize interaction terms with little regard for the definitions that distinguish them from other types of interactions (Caruson & MacManus, 2006; Mandell & Steelman, 2003; McNamara, 2012). Although practitioners seemingly understand that these terms have different meanings (Keast, Brown, & Mandell, 2007), this same level of understanding is not reflected in the literature where definition overlap runs rampant. Without acknowledging and defining each of these terms, researchers cannot properly consider the relative placement of the interaction on which they are focused. In addition, researchers seem to forego conveniently difficult discussions concerning definitions of interaction terms. The lack of a common language prevents reliable communication and collective understanding (Mandell & Steelman, 2003; McNamara, 2012). Conceptual advancement is stunted by definition overlap, ambiguity, and disregard.

Second, the appropriate application of each term cannot be acknowledged when they are used interchangeably. This blurs the boundaries between cooperation, coordination, and collaboration research. More specifically, much of the empirical research within the interaction literature focuses on preconditions, factors, triggers that influence a particular interaction, or purported factors for success (see, for example, Kuska, 2005; Ospina & Yaroni, 2003; Reilly, 2001). The difficulty with this line of inquiry is that many of the same factors are identified as influences to cooperation, coordination, and collaboration. While many of the same factors may influence each type of interaction, it seems likely that these factors will influence each type of interaction to different degrees. Since these distinctions are not made, there is confusion surrounding the optimal use of each interaction term (Keast, Mandell, & Brown, 2007). More needs to be done to better understand the contextual factors that help or hinder each type of interaction (McNamara, Miller-Stevens, & Morris, 2019).

Third, researchers often assume that the interaction of interest actually occurs (see, for example, Imperial, 2001; Kettl, 2003). The presence of a particular interaction term has not been tested empirically in relation to other types of interactions. Grounded theory is used by Imperial (2001) to identify factors that help or hinder collaboration during the development and implementation of six watershed programs. Collaboration is assumed to exist, and terms such as coordination and cooperation are not introduced or defined. It is disconcerting that this assumption is common within the interorganizational theory literature but not acknowledged.

Of the three interaction terms, it seems that collaboration is most emphasized within the current public administration literature. This emphasis focuses on factors that help or hinder collaboration (see, for example, Kuska, 2005; McNamara, Miller-Stevens, & Morris, 2019), collaborative management (see, for example, Agranoff, 2006; Bingham, Nabatchi, & O'Leary, 2005; Goldsmith & Eggers, 2004; McGuire, 2006; Meier & O'Toole, 2003; O'Leary, Choi, & Gerard, 2012), or the process of collaboration (see, for example, Bryson, Crosby, & Stone, 2006; Morris et al., 2013). However, literature's emphasis on collaboration between public organizations may not be warranted if public administrators tend to revert to coordinated interactions, as suggested by Keast, Brown, and Mandell (2007). Needless to say, it is important to differentiate empirically between the multiorganizational interactions that occur within the public sector. This differentiation requires a conceptual framework to highlight differences between cooperation, coordination, and collaboration (McNamara, 2012).

In order to enhance conceptual clarity and promote empirical progress, interorganizational theorists should look outside the literature. Multiorganizational interactions are also considered within the education literature (see, for example, Fagan, 1997; Goldman & Intriligator, 1990; Intriligator, 1994, 1992; LaRocco, 1997; Thatcher, 2007). The following section of this chapter describes the development of the Interorganizational Arrangement Model within the education literature.

Interorganizational Arrangement Model

The Interorganizational Arrangement Model (IAM) was originally developed in the health education literature (Goldman & Intriligator, 1990; Intriligator, 1992, 1994). It is used to examine interorganizational arrangements in settings such as health and human service delivery, medical and social service provision, and education (Fagan, 1997; Intriligator, 1992; LaRocco, 1997; Thatcher, 2007). More specifically, the IAM is a theoretical lens that can be used to explore relationships between organizations.

These relationships fall along a continuum of increased interdependence; cooperation, coordination, and collaboration are the interaction terms used to describe this continuum (Intriligator, 1992, 1994). The placement of these terms on the continuum is consistent with the interorganization theory literature. At one end of the spectrum, cooperation represents independent organizations who can individually accomplish the goals at hand. At the other end of the spectrum, collaboration represents interdependent organizations who must work together in order to accomplish the goals collectively identified. Coordination is placed in the middle of the two end points and represents organizations that require some assistance from other organizations in order to meet their individual goals. Placement on the continuum of interaction is measured by variables organized into the following four constructs: collective objective, collaborative infrastructure, collaborative procedures, and collaborative leadership (Thatcher, 2007). The relationships between the constructs of this model are identified in Figure 2.1.

According to the IAM, successful interorganizational efforts require use of the same type of interaction in all constructs (Intriligator, 1992). The type of interaction appropriate for the collective effort is determined by the interaction identified within the

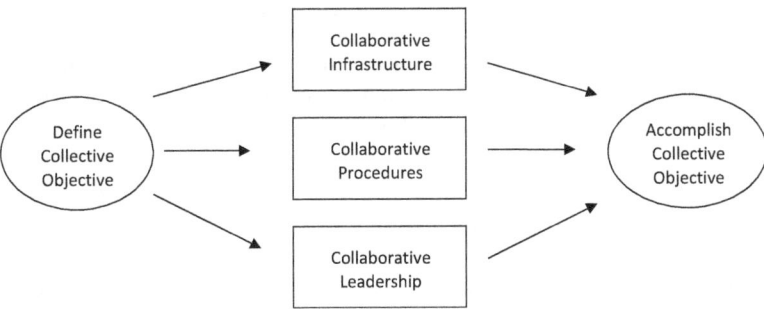

Figure 2.1 The Interorganizational Arrangement Model.

collective objective (Thatcher, 2007). The IAM assumes that inter-actions are not self-organizing. Instead, interagency planners must develop interactions within multiorganizational arrangements that are consistent with the collective objective (Thatcher, 2007).

The collective objective represents the goal that agencies work together to achieve (Intriligator, 1992, 1994). Four variables charac-terize the collective objective: time, complexity, single agency role, and the impetus for collective action (Thatcher, 2007). The level of interaction required by the collective objective drives the level of in-teraction desired throughout the remainder of the conceptual frame-work (Intriligator, 1994). Once the level of interaction required by the collective objective is identified, the level of interaction within the collaborative infrastructure, collaborative procedures, and collabo-rative leadership constructs are examined in order to categorize the type of interaction between organizations (Thatcher, 2007).

Collaborative infrastructure focuses on the organizational structures used to formalize and support relationships between organizations in the arrangement (Thatcher, 2007). Five variables characterize the construct of collaborative infrastructure: design, formality of the agreement, organizational autonomy, policy au-thority, and key personnel. Collaborative procedures are the pro-cesses developed to support operations within the arrangement (Thatcher, 2007). Five variables characterize the construct of collaborative procedures: information sharing, decision making, resolution of turf issues, resource allocation, and systems think-ing. Collaborative leadership is the ways in which behaviors of the member organizations support the arrangement (Thatcher, 2007). Five variables characterize the construct of collaborative leader-ship: incentives, commitment, trust, risk taking, and willingness to change.

Critique of the Interorganizational Arrangement Model

While the constructs and variables within the IAM provide a needed structure for deciphering between different interaction terms, there are five limitations that must be addressed. First, the model as-sumes that effective achievement of the collective objective requires all variables to operate at the same level of interdependence on the continuum (see, for example, Thatcher, 2007). Effective designs are those in which the type of interaction is aligned throughout all as-pects of the arrangement. This assertion is not supported by previ-ous research in which variation is prevalent. Based on the findings of Fagan (1997), LaRocco (1997), Olson (1996), and Thatcher (2007), it appears to be exceedingly difficult, if not unrealistic, to align empirically all variables within the same type of interaction. The IAM's assumption fails to consider that there may be differences in the relative importance of each variable in terms of achieving the interagency goal. If the relative importance of each factor varies, then an interagency arrangement may be effective even if all factors do not operate within the same type of interaction.

Second, the IAM assumes interorganizational interactions are formally and specifically planned based on the type of interaction necessitated by the collective objective (Goldman & Intriligator, 1990; LaRocco, 1997). After the type of interaction is determined by the collective objective, interagency planners strive to design an interorganizational arrangement aligned with this type of interaction. Success is based on the extent to which planners generate the conditions and relationships needed to link the appropriate interaction to the desired goal (Intriligator, 1992). In assuming that interactions are controlled by an elite group of planning personnel in a formal way, the model perpetuates an unwarranted emphasis on top-down implementation. Formal and specific planning may not be the impetus for all interactions that occur. The informal interactions that potentially occur in interorganizational settings are ignored despite Thatcher's (2007) development of the impetus for the collective action variable.

Third, nomenclature within the IAM biases it toward collaborative interactions. Thatcher (2007) places unnecessary emphasis on collaborative interactions by adding "collaborative" to nomenclature of all constructs. A model used to explore three different types of interaction terms should not emphasize one term to the detriment of others. By doing so, an inherent bias on collaboration as the desired end is created. While the IAM acknowledges that no one interaction is inherently better than another and each has the potential to be effective if used in the appropriate circumstance, the model's nomenclature defies this assumption. In addition, the leadership construct should be renamed entirely. Thatcher's (2007) definition of this construct suggests that a specific group of people oversee the arrangement. While oversight may be evident in coordinative interactions, it is not a component of cooperative or collaborative arrangements. Interorganizational cooperation typically occurs as relationships form between lower levels of organizational structures. While leaders can emerge during the early stages of interorganizational collaboration, the interaction is sustained by partners holding positions of equal authority. Therefore, the definition's emphasis on oversight is misplaced.

Fourth, operationalizations within the IAM are unclear. Ambiguity makes it difficult to apply the IAM to other research settings, because linking the data collected to the model's current operationalizations is challenging. In order to expand the application of this model to other settings, the operationalizations must be further clarified and supported by the literature.

Fifth, the current model inaccurately portrays collaboration as an arrangement in which participants align their interests with the new organizational arrangement without consideration for the interests of their individual organizations. This is inconsistent with the broader collaboration literature, which portrays this type of interaction in a far more complicated manner than acknowledged by the model. According to the public administration literature, a commitment to the collective objective does not decrease stakeholders' commitments to their individual organizations (Keast et al., 2004). Organizations often struggle with dual commitments to their individual organizations and the multiorganizational arrangement (O'Leary & Bingham, 2007b). In fact, collaboration occurs when stakeholders work together to solve one another's interests without giving up any of their own (Wood & Gray, 1991). This inconsistency between the education and interorganizational theory literatures is addressed in the next section while the complex nature of collaboration is acknowledged.

Ambiguity in the original model's design, assumptions, construct nomenclature, and operationalizations are problematic. Extensive changes to this model improve its transferability to other areas. These changes are further explained in the following section as the model transforms into the Multiorganizational Interaction Model.

The Multiorganizational Interaction Model

The Multiorganizational Interaction Model (MIM) provides the theoretical basis for this study. This section begins by explaining the transformation leading to the development of this model. Ambiguities from an earlier model are eliminated by significantly altering its design, assumptions, construct nomenclature, and operationalizations. The terms used to describe interactions within the MIM continue to represent a continuum of increased interaction and are identified as cooperation, coordination, and collaboration (see Keast, Brown, & Mandell, 2007; McNamara, 2012; Thomson & Perry, 2006). The definitions of these terms align with the policy implementation and interorganizational theory literatures.

The design of the MIM suggests that all four of the model's constructs simultaneously impact the interaction continuum. Placement along the continuum of interaction is measured by variables organized into the following four constructs: interorganizational policy objective, interorganizational infrastructure, interorganizational procedures, and organizational management. Relationships between the model's constructs and the continuum are illustrated in Figure 2.2.

Rather than assuming that the policy objective drives the type of interaction present in the other constructs, each construct is assumed to impact the continuum of interaction independently and collectively. Implementation of the interorganizational policy objective does not require all constructs to operate at the same level of interdependence. This eliminates the education literature's assumption that effective interactions are formally planned to align with the type of interaction identified in the collective objective. By acknowledging that formal planning may not be the impetus for all interactions, there is room within the MIM to account for informal interactions. While not pictorially expressed, the relative importance of the model's variables may differ in their impact on interactions during multiorganizational implementation. Discussion during interviews enhances meaning and provides context regarding the impact of the model's variables.

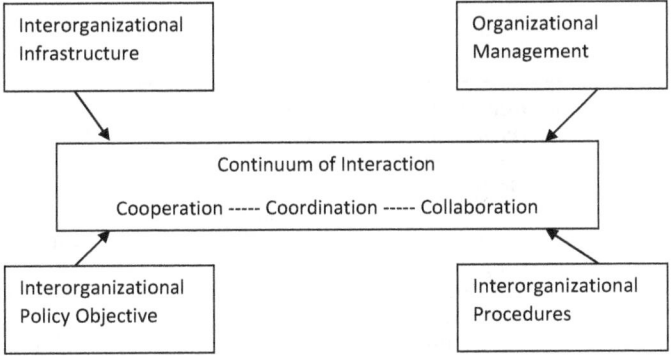

Figure 2.2 The Multiorganizational Interaction Model.

Changes to construct nomenclature eliminate an inherent bias toward collaboration. By replacing "collaborative" with "interorganizational" in the construct nomenclature, the MIM acknowledges the potential for one of three interactions to take place. This change supports the model's theoretical focus on distinguishing different types of interactions. An additional nomenclature change is made to replace the collaborative leadership construct with a construct called organizational management. Through this change, the MIM acknowledges that arrangements, depending on the type of interaction, may be managed in different ways. Therefore, this model does not assume that all types of interactions are overseen by formally identified personnel.

Operationalizations of the MIM constructs are developed through alignment with the policy implementation and interorganizational theory literatures. These revisions eliminate ambiguity and helps us clearly distinguish between the three types of interactions. In addition, these operationalizations acknowledge the complexities of collaborative interactions expressed in the interorganizational theory literature. Operationalizations for each interaction term are used to determine a partnerships overall placement along the continuum. The variables used to operationalize the model's constructs are identified in Table 2.1. While Thatcher (2007) utilizes the same variable nomenclature, construct nomenclature is significantly different.

The remainder of this section focuses on linking the policy implementation and interorganizational theory literatures to the operationalizations of the MIM.

Table 2.1 Constructs and Variables of the Multiorganizational Interaction Model

Interorganizational Policy Objective	Interorganizational Infrastructure	Interorganizational Procedures	Organizational Management
Time	Design	Information sharing	Incentives
Difficulty	Formality of the agreement	Decision making	Commitment
Role of single organization	Organizational autonomy	Resolution of turf issues	Trust
Impetus for collective action	Policy authority	Resource allocation	Risk taking
	Key personnel	Systems thinking	Willingness to change

Interorganizational Policy Objective Construct

The interorganizational policy objective represents a policy-mandated goal that organizations work together to achieve (Caruson & MacManus, 2006; O'Toole & Montjoy, 1984; Raelin, 1980, 1982). Four variables characterize the construct of interorganizational policy objective: time, difficulty, role of single organization, and the impetus for collective action (Thatcher, 2007). Table 2.2 displays the operationalizations of each variable within the interorganizational policy objective construct and places them along the continuum of interaction. These revisions align with the use of cooperation, coordination, and collaboration in the policy implementation and interorganizational theory literatures. Theoretical support for these operationalizations is provided.

In a cooperative interaction, independent organizations voluntarily work together to pursue a short-term goal involving relatively simple tasks (O'Leary & Bingham, 2007a; Thomas, 1997). Although it is possible to achieve organizational goals by working alone, actors within each organization make a deliberate decision to work together because it is helpful to their world of work (Kuska, 2005; O'Leary & Bingham, 2007a; Rogers & Whetten, 1982). This decision is informal and based on recognized opportunities to share information, build capacity, or generate synergy that serves individual organizations (May, 1995; Thomas, 1997). The desire to work together may be triggered by changes in external factors and the desire to avoid negative impacts (Ospina & Yaroni, 2003).

In a coordinated interaction, organizations establish formal relationships to pursue longer-term goals aligned with repeatable tasks deemed to be compatible with each individual organization's interests (Jennings, 1994; Keast, Brown, & Mandell, 2007; Mandell, 1994; Schlossberg, 2004). Organizations are semiautonomous and some outside assistance is needed to accomplish organizational goals. This notion is supported by the interviews conducted by Keast, Brown, and Mandell (2007), in which a majority of respondents indicate that coordination is used to "drive" a particular initiative or outcome. These relationships are typically formed for two reasons. First, organizations may be mandated to work together (see, for example, Hall & O'Toole, 2004; Mandell, 1994; McNamara, 2016a; O'Toole, 1983, 1991, 1995). Coordination can be mandated by government to reorganize, minimize duplication, minimize conflict, or promote cohesion (Boston, 1992; McNamara, Morris, & Mayer, 2014). For example, the Department of Homeland Security was established to support coordinated intergovernmental responses to

Table 2.2 Variable Operationalizations: Interorganizational Policy Objective Construct

Variable	Cooperation	Coordination	Collaboration
Time	Short-term	Longer-term	Long-term, evolutionary nature.
Difficulty	Simple task	Multifaceted tasks, repeatable	Complex tasks that are highly varied and diverse; or situations of crisis.
Role of single organizations	Organizations are independent; it is possible for them to accomplish the task individually.	Organizations require some assistance from other organizations to accomplish individual goals/missions.	Organizations are interdependent; each organization is one element of the larger system.
Impetus for collective action	Typically voluntary, organizations initiate collective action because it is helpful to their world of work and it builds capacity that serves the individual organization. Changes in external factors trigger organizations to search for new solutions.	Voluntary or mandated, linkages are mobilized because compatible mission areas mutually increase abilities to achieve individual goals. An interagency liaison or boundary spanner may forge these relationships to meet resource needs or shared interests. Legislative mandate or grant contracts may enhance cohesion or minimize duplication.	Voluntary or mandated, organizations with mutual or complementary interests come together because they cannot achieve the desired goal or address the identified problem without working together. Organizations share responsibility for tasks that are interconnected or cannot be accomplished individually. A lead agency or convening organization brings relevant stakeholders together and legitimizes collective action.

national disasters, enhance role clarity, and minimize duplication (Wise & Nader, 2002).

Second, organizations may work together in a coordinative relationship based on increased abilities to achieve individual goals due to compatible mission areas (Jennings & Krane, 1994; Kettl, 2003; Van de Ven & Walker, 1984). In Jennings' (1994) study on state- and local-level employment and training programs, 26% of respondents indicate that shared role definitions are an important factor in forming coordination. Domain similarity may facilitate relationships between organizations based on complementary resources and shared professional skills (Van de Ven & Walker, 1984). Therefore, a coordinative interaction is mutually beneficial (Peters, 1998; Schlossberg, 2004). Linkages of exchange are typically established between the organizations (Jennings, 1994; Jennings & Ewalt, 1998); a boundary spanner, interagency liaison, or facilitator may help forge these linkages (Kapucu, 2006; Thompson, 1967; Van de Ven & Walker, 1984). In their research pertaining to the employment and training services under the Job Training Partnership Act, Jennings and Ewalt (1998) conclude that 71% of respondents use interdepartmental liaisons as a coordination technique. This finding is consistent with Jennings' earlier (1994) work.

In a collaborative interaction, organizations establish highly interdependent relationships that evolve as organizations interact with one another to attain long-term goals (Huxham, 2003; O'Leary & Bingham, 2007a; Thomson & Perry, 2006). Interdependence develops as organizations share responsibility for highly complex problems or crisis that prevents them from acting alone (Bryson, Crosby, & Stone, 2006; Gray, 1985; Morris et al., 2013). Each organization is considered an essential element of the larger interdependent system (McNamara, 2012).

Collaborative membership can be voluntary or mandated (Agranoff, 2006; Imperial, 2005, McNamara, 2016a). While a contractual arrangement or statutory action may be used to bring participating organizations together, interpersonal relationships must go beyond the terms of the contract (Keast et al., 2004; Mandell, 1994). The extent of the relationship can change based on the environment or the extent to which collective action is reciprocated among partnering organizations (Thomson & Perry, 2006). If multiorganizational implementation does not occur as expected, organizations may renegotiate their commitment to the arrangement (Mattessich, Murray-Close, & Monsey, 2001). These changes will reshape the dynamics in the interaction (Mandell & Steelman, 2003). Since collaboration is time-consuming and costly, this interaction should only be used when addressing a highly complex problem, in times of crisis, or when other forms of interaction will not suffice (Bryson, Crosby, & Stone, 2006; Gray, 1985; Keast et al., 2004).

A convening or referent organization plays a significant role in establishing the collaborative (McNamara, 2016b; McNamara, Leavitt, & Morris, 2008; Wood & Gray, 1991). This organization legitimizes the arrangement by identifying an important problem and bringing relevant stakeholders together to address a particular purpose (Wood & Gray, 1991). Through resource and information exchange, a referent organization facilitates interactions between organizations and generates stability within the organizational environment (Morris & Burns, 1997). According to Gray (1989), a referent organization must have the following: power to persuade stakeholder participation, credibility among stakeholders, abilities to establish collaborative processes, and capabilities to identify relevant stakeholders. Since personnel within the convening organization do not have the formal authority over other organizations within the collaborative network, informal influence must be generated through expertise and credibility (Gray, 1989; Morris et al., 2013). It is essential that all members of the collaboration perceive the convener to hold legitimate authority to organize the arrangement (Gray, 1985; McNamara & Morris, 2012).

Interorganizational Infrastructure Construct

Interorganizational infrastructure focuses on the ways in which organizations within the multiorganizational arrangement generate and structure relationships. Five variables characterize the construct of interorganizational infrastructure: design, formality of the agreement, organizational autonomy, policy authority, and key personnel (Thatcher, 2007). Table 2.3 displays the revised operationalizations of each variable within the interorganizational infrastructure construct and places them along the continuum of interaction. These revisions are aligned with the use of cooperation, coordination, and collaboration within the public administration and organization theory literatures. Theoretical support for these operationalizations is provided.

In a cooperative interaction, a commonly defined structure does not exist because organizations work within their existing organizational structures (Keast, Brown, & Mandell, 2007; McNamara, 2012). The simplicity of tasks does not require an interagency

Table 2.3 Variable Operationalizations: Interorganizational Infrastructure Construct

Variable	Cooperation	Coordination	Collaboration
Design	Individuals work independently within existing organizational structures; an interagency staff is unnecessary.	Each organization's hierarchical structure is used to centrally manage specialized roles and responsibilities. Centralization may involve reorganization or consolidation of programs/activities.	Partner organizations jointly develop shared power arrangements to support mutually beneficial interests. New program structures are developed based on the needs of a specific policy/goal. An administrative staff is present to sustain collective efforts.
Formality of the agreement	Individual organizations informally agree to work together to achieve individual goals.	Mechanisms, such as contractual or nonfinancial agreements, formalize relationships between organizations. Agreements, clearly identifying each organization's roles and responsibilities, are often developed and/or reviewed by a higher authority.	Key stakeholders jointly draft a shared purpose and develop a course of action based on mutually agreed upon roles and responsibilities, rules, goals, and organizational boundaries.
Organizational autonomy	Organizations are fully autonomous.	Organizations are semiautonomous; individual organizations require some assistance from other organizations to achieve goals.	Organizations are not autonomous; operations within organizations are intertwined.
Policy authority	No multiorganizational policy decisions are made. Preexisting policies, established by the individual organizations, are followed.	Organizations maintain individual authority over the policies that govern their respective organizations. Policies pertaining to coordinated efforts may be developed, but they are compatible with the policies already established within the individual organizations.	Partner organizations jointly develop policies and procedures that govern the collective group. Multiorganizational policies and procedures include working rules that specify which stakeholders can make decisions, who will guide collective actions, and the distribution of costs/benefits.
Key personnel	Organizational leadership is not involved in decisions to work together.	There is a distinction between leaders and managers; leaders make decisions while managers implement and administer these decisions. A facilitator may be identified to coordinate actions at the local level.	Although no one is typically in charge, a lead organization may propose policies/rules to which the collective group must mutually agree to implement. Membership, role definitions, and responsibilities adapt to the task at hand. Each role is considered equally important.

staff, a multiorganizational structure, or a collective planning effort (Reilly, 2001). Organizations informally work together based on a mutual benefit where individual organizational interests are emphasized (Keast, Brown, & Mandell, 2007). Therefore, organizations retain separate entities, maintain individual control of resources, and make independent policy decisions (Keast, Brown, & Mandell, 2007). Preexisting policies, established by the individual organizations are utilized and remain unchanged (Keast, Brown, & Mandell, 2007). Since cooperative interactions usually occur at the lower levels of an organization and involve few resources (2007), leaders from individual organizations are not typically involved in decisions to work together.

In a coordinated interaction, formal relationships are emphasized within a hierarchical structure to specialize roles and responsibilities between otherwise independent organizations (Mandell, 1994; McNamara, 2016a; Reilly, 2001). Administrative efforts are centrally controlled and local-level planning is minimized to maintain control over organizational components (Jennings & Ewalt, 1998). Centralization may involve reorganization or consolidation to minimize duplication (Boston, 1992; Jennings, 1994). Relationships between organizations are often formalized through contractual or nonfinancial agreements (Jennings & Krane, 1994). In their research of the Job Opportunities and Basic Skills Program, Jennings and Krane (1994) conclude that organizations enter into contractual or nonfinancial agreements to leverage resources and maximize the number of clients receiving services. Contractual agreements between local agencies establish areas for joint action and outline roles and responsibilities; these contracts are reviewed by a regional authority representative of local organizations (Jennings & Krane, 1994).

Organizations participating in coordinated activities are considered semiautonomous (Keast, Brown, & Mandell, 2007). They maintain individual authority over the policies that govern their respective organizations, but they agree to participate in some specific collective activity (Keast, Brown, & Mandell, 2007). Since organizations focus on their individual missions (Jennings, 1994), policies established for the collective arrangement must be congruent with those of the individual organizations. The leaders within each organization make their own policy decisions and managers are expected to implement these decisions (Thomson & Perry, 2006). A liaison may be designated to facilitate and guide interactions between organizations at the local level (Jennings & Ewalt,

1998; Schlossberg, 2004). These boundary spanners can be used to broker relationships and identify areas of interdependence (Keast et al., 2004).

In a collaborative interaction, organizations jointly develop a structure of shared power to address collective interests (McNamara, 2012). While this structure can take many forms, it is important to recognize that individual organizations coexist within a new program structure (Mandell, 1994; Mattessich, Murray-Close, & Monsey, 2001). Administrative staff may be used to bring organizations together and implement collective policies (Thomson & Perry, 2006). In Agranoff's (2006) research on public management networks, a staff element is present in all of the networks analyzed. These actors are involved in all operations of the collaboration.

It is important for partnering organizations to understand their roles and responsibilities (Reilly, 2001; Thomson & Perry, 2006). In order for this to occur, key stakeholders must establish shared rules, develop a collective purpose, and jointly decide on a course of action (Imperial, 2000). As a result, organizations relinquish some autonomy to the collective unit. Formal and informal agreements can be used among partnering organizations, and the benefits of both should be considered (Bryson, Crosby, & Stone, 2006; Imperial, 2005; Thomson & Perry, 2006). Informal agreements may easily support the evolutionary nature of collaborations; changes are made as the arrangement grows, partners change, or the problem domain shifts (Bryson, Crosby, & Stone, 2006). On the other hand, stability can be created by formalizing the social norms and agreements that establish over time (Morris, McNamara, & Belcher, 2019). Multiorganizational policies generate working rules to specify which stakeholders can make decisions, who will guide collective actions, and the distribution of costs or benefits.

While no one is typically in charge of a collaborative arrangement, this does not mean that all organizations are always of equal status (Keast et al., 2004). Often, policy mandates or agency rules identify a lead organization responsible for implementation. While other organizations are not under the control of the lead agency, this organization can propose formal rules for the collective group to consider (Keast et al., 2004). Even when certain organizations are more influential than others, they recognize that combined efforts are needed to accomplish the objective (Mandell, 1994).

Dynamics surrounding the collaborative arrangement are in constant flux (Huxham, 2003). Therefore, changes to membership may be required and participants must develop competencies for multiple roles (Gray, 1985; Morris, McNamara, & Belcher, 2019). Despite the needed flexibility among personnel, it is important for the collaborative group to identify sponsors and champions among its formal and informal leaders. Sponsors provide authority and resources to legitimize the collaboration while champions have the expertise to sustain daily operations (Agranoff, 2006; Mandell & Steelman, 2003). In Agranoff's (2006) research on public management networks, champions play a significant role in encouraging other organizations to support the collaborative arrangement. The collaboration literature also refers to the strategic leader of the arrangement as collaborative entrepreneur, convener, or manager (McNamara, 2016b; McNamara & Morris, 2012).

Interorganizational Procedures Construct

Interorganizational procedures are the processes developed to support operations or sustain relationships within the multiorganizational arrangement. Five variables characterize the construct of interorganizational procedures: information sharing, decision making, resolution of turf issues, resource allocation, and systems thinking (Thatcher, 2007). Table 2.4 displays the revised operationalizations of each variable within the interorganizational procedures construct and places them along the continuum of interaction. These revisions are aligned with the use of cooperation, coordination, and collaboration within the public administration and organization theory literatures. Theoretical support for these operationalizations is provided.

In a cooperative interaction, communication is an important factor in developing and sustaining the arrangement. An emphasis is placed on sharing information to create harmonized efforts (Nylen, 2007). Dialogue is maintained through informal communication channels to allow information sharing among participants (Ospina & Yaroni, 2003). While basic information is initially shared, continuous dialogue creates opportunities for discussing a wider range of topics (2003). Since each organization retains their autonomy (Keast, Brown, & Mandell, 2007), joint decision-making processes are not developed. Instead, organizational systems remain unchanged and operational decisions are independently made by each organization. Turf issues are avoided, and organizational systems remain independent. Discretionary funds may be used in pursuit of individual goals; resources are not pooled. Units of exchange are determined at the lowest possible level.

Table 2.4 Variable Operationalizations: Interorganizational Procedures Construct

Variable	Cooperation	Coordination	Collaboration
Information sharing	Dialogue is maintained through informal relationships between participants. Basic information is initially shared. Continuous dialogue creates opportunities for discussing a wider range of topics.	Formal and informal communication channels are used to link vertical and horizontal organizational levels.	Emphasize open and frequent communications between partners to reduce information asymmetries. Formal and informal channels are used to widely disseminate information concerning the collective group. Understanding enhanced by a willingness to share information about individual organizations and what can/cannot be offered to the collective group.
Decision making	Decisions are made independently; rules that guide collective decision making are not necessary.	Centralized decision making is practiced; a lead organization(s) dominates the decision-making process.	Participative decision making based on consensus and compromise, generates rules to govern activities and relationships between organizations. Representatives have latitude to negotiate rules and deliberate agreements to identify common ground.
Resolution of turf issues	Turf issues between participating organizations are avoided based on organizational tendencies to function independently	A neutral facilitator, outside convener, or full-time coordinator is employed to resolve turf issues.	Conflicting roles based on incongruent demands from individual organization and group. Consider adjusting policies and procedures to reduce conflict while maximizing common ground.
Resource allocation	Discretionary funds may be used in the pursuit of individual goals. Resources are not pooled. Units of exchange are determined at the lowest possible level.	Organizations exchange resources to increase each organization's abilities to achieve individual goals. Mandates or grant arrangements may provide resources. Resource needs may be satisfied by a preexisting program within an individual organization.	Pooled resources; allocation is based on balancing evolving needs of the collective group with individual constraints. Individual organizations have resources, skills, or knowledge needed to achieve collective goal. Organizational resources are allocated to support the activities of the collective unit.
Systems thinking	Organizational systems remain unchanged.	Compatible information systems can enhance coordination.	Databases are integrated to create linkages and share information between multiple layers of partner organizations.

In a coordinated interaction, formal and informal communication channels are used to exchange information within and across organizational boundaries (Boston, 1992; Jennings, 1994). In Jennings' (1994) study of employment and training activities in the Job Training Partnership Act, the following formal information channels are identified: working partnerships, regular meetings of staff from different units, and interdepartmental liaisons. Sixty percent of respondents indicate that working partnerships and regular meetings of staff from different units contribute very much to coordination; more than 50% of respondents indicate that interdepartmental liaisons contribute very much to coordination (Jennings, 1994).

Some of the informal information channels identified include workshops or common geographical boundaries. Seventy-eight percent of respondents indicate that they use workshops as a coordination technique (Jennings & Ewalt, 1998). Based on increased opportunities for interaction and information sharing, Jennings and Krane (1994) determine that collocation enhances coordination. Interorganizational interactions can occur between the top and lower levels of multiple organizations (Kapucu, 2006). Therefore, formal and informal communication channels are important to linking vertical and horizontal organizational levels. Organizational systems can be designed to facilitate increased communication between organizations. Incompatible organizational systems may prevent coordination (Jennings & Krane, 1994).

The decision-making process is highly centralized and dominated by the lead agency (Jennings & Krane, 1994). Issues related to turf can be barriers to coordination (Jennings & Ewalt, 1998). A neutral facilitator, separate from a boundary spanner or convener, can be used to help organizations recognize that they may have different goals that should be respected by the collective group. This coordinator, whether full-time or part-time, should be dedicated to resolving conflicts (Schlossberg, 2004). In studies conducted by Jennings and Ewalt (1998) and Jennings and Krane (1994), resource limitations are also identified as a potential barrier to coordination. Resources are the physical property and financial assets used to achieve the organization's missions (Jennings, 1994; Jennings & Krane, 1994). For example, Jennings (1994) concludes that more than 50% of respondents indicate that joint program funding contributes very much to coordination. Resource exchange is often an important element of coordination (Van de Ven & Walker, 1984). Organizations may combine complementary resources to create mutually beneficial relationships that enhance each organization's abilities to achieve individual goals (Peters, 1998; Schlossberg, 2004). Although it is unlikely, policy mandates may include a provision for joint resources (O'Toole & Montjoy, 1984).

In collaborative interactions, an emphasis is placed on open and frequent communications between partners (Thomson & Perry, 2006). Information pertaining to individual organizations and the collective arrangement can be passed through formal channels or informally through personnel interactions (McNamara, 2016a). When information is shared amongst all partners, a common knowledge base is built to promote understanding (Imperial, 2001). In addition, sharing information can reduce information asymmetries provided that organizations clearly state what they can and cannot offer the multiorganizational arrangement.

Decisions regarding the arrangement's direction and operations are made collectively (Mandell, 1994). Joint decision making relies on consensus and compromise to bridge differences among individual organizations (Agranoff, 2006; Mandell, 1999; Reilly, 2001). In order to identify common ground, representatives of individual organizations must have discretion to negotiate rules and make organizational decisions based on group deliberation (Mattessich, Murray-Close, & Monsey, 2001). Discretion involves the use of personal judgment to make a decision between multiple alternatives (Carrington, 2005). Turf issues may arise as each organization is placed in conflicting roles based on incongruent demands from the individual and collective organizations (McNamara, Miller-Stevens, & Morris, 2019; Thomson & Perry, 2006). Considerations are given to adjusting policies and procedures to reduce conflict (Mattessich, Murray-Close, & Monsey, 2001). In Agranoff's (2006) research on 14 collaborative networks involving managers in federal, state, and local governments, problems related to turf do not significantly impact the network.

An organization's abilities to collaborate can be hindered by a lack of staffing, funding, or expertise (Morris, McNamara, & Belcher, 2019). Therefore, the presence of consistent financial and personnel support is important (Imperial, 2001). To overcome these limitations, organizations often pool their resources (Gray, 1985; Nylen, 2007). For example, Imperial's (2005) research indicates that resources are pooled in various ways ranging from informal sharing of equipment to more formal methods of combining financial resources and collocating staffing units. In addition, Agranoff's (2006) research on collaborative networks indicates that government managers are willing to pool resources. However, decisions are difficult to make when some agencies are unwilling to contribute to the pool of resources (Agranoff, 2006).

Interorganizational systems are developed to involve personnel from multiple layers within partnering organizations (Mattessich, Murray-Close, & Monsey, 2001). It is important to involve people representing a variety of organizational positions. In his study of watersheds, Imperial (2005) finds that interagency databases, joint research, and joint technical information are used to share knowledge during collaborative activities.

Organizational Management Construct

Organizational management pertains to the way in which behaviors within and between member organizations support the multiorganizational arrangement (McNamara, 2012). Five variables characterize the construct of organizational management: incentives, commitment, trust, risk taking, and willingness to change (Thatcher, 2007). Table 2.5 displays the revised operationalizations of each variable within the organizational management construct and places them along the continuum of interaction. These revisions are aligned with the use of cooperation, coordination, and collaboration within the public administration and organization theory literatures. Theoretical support for these operationalizations is provided.

In cooperative interactions, incentives to interact are based on recognizing opportunities for synergistic benefits (Thomas, 1997). Cooperative behavior may be triggered by changes in external factors and an organization's desire to identify common ground in hopes of avoiding negative impacts (Ospina & Yaroni, 2003). When engaging in cooperative interaction, organizations do not need to make any changes to their current operations or missions (Keast, Brown, & Mandell, 2007; McNamara, 2012). Since interests of the individual organizations remain paramount, the cooperative effort is aligned with each organization's standard operating procedures. Organizations maintain complete independence in establishing the rules, roles and responsibilities, and policies that govern their involvement with the collective action (Keast, Brown, & Mandell, 2007). Trust relationships are not required but can develop when organizations consistently share honest information (Ospina & Yaroni, 2003). Since organizational policies and operations remain

Table 2.5 Variable Operationalizations: Organizational Management Construct

Variable	Cooperation	Coordination	Collaboration
Incentives	Opportunities for synergistic benefits are realized based on the desire to avoid negative impacts resulting from changes in external factors.	Statutes or grant contracts may provide funding or resource incentives to support the collective effort. Leaders identify benefits in working together and emphasize the importance of these benefits to subordinates.	Incentives are provided by the collective group and individual organizations to encourage individuals to stay involved in the collective effort.
Commitment	Work is completed as part of the regular job responsibilities conducted within the individual organization. Interests of the individual organization remain paramount. Colleagues may encourage each other to work with personnel in other organizations.	A supervisory administrative body actively encourages organizations to work together. Linkages between organizations are recognized when benefits are perceived to outweigh the costs.	Members are committed to intra- and interorganizational partners; collective interests must constantly be balanced with self-interest. Participation is justified by perceptions that the collective interest serves each organization's interests. Mutual commitment expands as organizations reciprocate collective action.
Willingness to change	Participating organizations maintain complete independence in establishing the rules, roles and responsibilities, and policies to govern the organization. An organization's standard operating procedures are not modified.	Leaders explore modifications to standard operating procedures when supporting operational goals aligned with individual organizational missions.	Partner organizations mutually adjust to the rules, roles and responsibilities, and policies collectively established to govern the collective unit. Changes to an organization's standard operating procedures are considered when needed to align with those of the collective unit.
Trust	Trust relationships are not required but can develop when organizations consistently share honest information.	Leaders work closely to create relationships based on trust.	Trust between organizations is necessary. Partners reinforce trust in each other by sharing information through open communication. A history of supportive interactions sustains and legitimizes relationships.
Risk taking	Organizations do not generally engage in risk-taking behavior. Characterized by low levels of risk.	Some interdependencies may be formed based on resource needs. Characterized by moderate levels of risk.	Integrated approaches develop and create dependency. Adherence to shared policies may require organizations to depart from normal behavior. Characterized by high levels of risk.

independent, cooperative relationships are considered to have low levels of risk (Keast, Brown, & Mandell, 2007).

In a coordinated interaction, incentives to participate in the collective action can come from two sources. First, statutory provisions may provide funding incentives to support the collective effort (Jennings, 1994; McNamara, 2012). Provisions within a grant contract may designate an interorganizational liaison and require organizations to work together to implement the goals of the grant (Schlossberg, 2004). Second, leaders within the hierarchical structure may identify a benefit in working together and communicate the importance of coordination to their subordinates (Jennings, 1994; Schlossberg, 2004). In Jennings (1994), 70% of administrators indicate that leaders play an important role in establishing commitment to coordination. The importance placed on support from leadership is significantly higher than the importance placed on financial sanctions, interpersonal relations, or interagency recognition. Therefore, an organization's supervisory level must be actively involved to encourage coordination.

Organizations are committed to the relationship provided that their interests outweigh the costs. Partnerships between member organizations are formalized to ensure clear role definition (Jennings, 1994). Although individual organizations retain authority over decision making, joint planning and information sharing does occur (Keast, Brown, & Mandell, 2007). As a result of these processes, modifications to an individual organization's standard operating procedures may be explored when it is in support of its mission. An emphasis is placed on developing trust among the leaders in charge of the individual organizations (Jennings & Krane, 1994; McNamara, Morris, & Mayer, 2014). Since interdependencies can form on the basis of resource exchange, coordinated relationships are considered to have moderate levels of risk (Keast, Brown, & Mandell, 2007).

In a collaborative interaction, incentives are provided by the collective and individual organizations to encourage individuals to stay involved with the collaborative effort (Mattessich, Murray-Close, & Monsey, 2001). Organizations and individuals should be rewarded for their participation in the collaborative effort (Imperial, 2005; Gray, 1985). Imperial's (2005) research on collaboration within watershed programs indicates that the presence of rewards provides incentives for organizations to continue to work together. The effects of the incentives must be monitored to ensure they motivate members as intended (Mattessich, Murray-Close, & Monsey, 2001).

Power imbalances and competing organizational interests may create tension and conflict among members of the collaborative arrangement (Mandell & Steelman, 2003; McNamara, Miller-Stevens, & Morris, 2019). It is important to recognize each participant's struggle to meet the commitments of the interorganizational unit while also supporting the potentially conflicting interests of their individual organizations (O'Leary & Bingham, 2007b; Thomson & Perry, 2006). The term "program rationale" is used to describe the mindset toward the collective arrangement as a legitimate influence over individual behavior (Mandell, 1994). While this commitment to the collective arrangement plays an important role in facilitating joint agreements (Mandell, 1994), it does not decrease members' commitments to their individual organizations (Keast et al., 2004). Despite potentially different interests, collaboration typically occurs when stakeholders work together to resolve inherently complex problems without giving up any of their own interests (Wood & Gray, 1991). The extent to which individual organizational interests are met will determine an organization's willingness to support and commit to the collective endeavor (Imperial, 2001). Participation is justified based on a perception that the collective interest also serves an organization's individual interests (Thomson & Perry, 2006).

In order for a collaborative arrangement to accomplish the desired objective, two or more organizations must mutually adjust to collective policies and procedures (Mandell, 1999). Partner organizations become interdependent as integrated policies and operations are established (Imperial, 2000, 2005). Changes to an organization's standard operating procedures are made when needed to align with those jointly developed by the organizations within the collective unit. It can be challenging to establish policies and procedures for the entire group to follow, because group membership is comprised of many different organizations with different views (Mandell & Steelman, 2003). When members are unable to accommodate other organizations, a collaborative arrangement will not be established (Mandell & Steelman, 2003; McNamara, Miller-Stevens, & Morris, 2019). The history of interactions between organizations will affect the extent to which organizations are willing to change (Mandell & Steelman, 2003). As participants gain a broader view of their relationships with other organizations, some of the bureaucratic boundaries preventing change may dissolve (Keast et al., 2004).

It is through this process that a new value set can be established to change the views of individual participants (Keast et al., 2004).

The likelihood that collective action will occur increases when members have a reputation for trustworthiness (Huxham, 2003; Morris, et al., 2013). It is important for organizations within the arrangement to believe that partners are committed to the collective objective, will act within the established rules, and honestly negotiate with other organizations. This trust is also referred to as an "ethic of collaboration" (Thomson & Perry, 2006, p. 25). While trust is a critical component of collaborations, it takes time and resources to develop and sustain (McNamara, 2012). Prior relationships and open communications may help partners develop mutual understanding and reduce vulnerability (Huxham, 2003; Mandell & Steelman, 2003). Trust facilitates sustained relationships between partners as an emphasis on formal organizational roles and contractual arrangements diminish (Bryson, Crosby, & Stone, 2006; Thomson & Perry, 2006). Collaborative arrangements involve high levels of risk (Keast et al., 2004). As partner organizations develop integrated policies and operations, dependency between organizations is created (Imperial, 2000, 2005).

Summary

Increased levels of interaction are not inherently more desirable, and a specific interaction will not be effective in all settings. Costs, challenges, and risks increase as interactions proceed along the continuum (Nylen, 2007; Schlossberg, 2004). Since there are several types of interactions to consider, the appropriate interaction for a particular situation must be carefully chosen.

This study uses the Multiorganizational Interaction Model to explore the types of interactions used in multiorganizational arrangements. This model distinguishes between interaction terms and aligns operationalizations with the policy implementation and interorganizational theory literatures. The operationalizations of these variables are critical to placing multiorganizational relationships on the continuum of interaction. Application of the MIM allows for empirical testing of operationalizations. Its use in a policy implementation setting will help determine the extent to which the model can be used to explain interactions between organizations. In addition, a potential for informal interactions between organizations is acknowledged by assuming that variables within each of the four constructs simultaneously impact the continuum of interaction. The MIM guides the development of research questions for this study.

References

Agranoff, R. (2006). Inside collaborative networks: Ten lessons for public managers. *Public Administration Review*, 66. 56–65.

Alexander, E. (1989). Improbably implementation: The Pressman-Wildavsky paradox revisited. *Journal of Public Policy*, 9(4). 451–466.

Althaus, R., & Yarwood, D. (1993). Organizational domain overlap with cooperative outcomes: The Departments of Agriculture and State and International Agricultural Policy during the Carter Administration. *Public Administration Review*, 53(4). 357–367.

Ansel, C., & Gash, A. (2008). Collaborative governance in theory and in practice. *Journal of Public Administration, Research and Theory*, 18(4). 543–571.

Bingham, L., Nabatchi, T., & O'Leary, R. (2005). The new governance: Practices and processes for stakeholder and citizen participation in the work of government. *Public Administration Review*, 65(5). 547–558.

Boston, J. (1992). The problems of policy coordination: The New Zealand experience. *Governance: An International Journal of Policy and Administration*, 5(1). 88–103.

Bovens, M., & Zouridis, S. (2002). From street-level to system-level bureaucracies: How information and communication technology is transforming administrative discretion and constitutional control. *Public Administration Review*, 62(2). 174–184.

Brummel, R., Nelson, K., & Jakes, P. (2012). Burning through organizational boundaries? Examining inter-organizational communication networks in policy-mandated collaborative bushfire planning groups. *Global Environmental Change*, 22. 516–528.

Bryson, J., Crosby, B., & Stone, M. (2006). The design and implementation of cross-sector collaborations: Propositions from the literature. *Public Administration Review*, 66. 44–55.

Callahan, R. (2007). Governance: The collision of politics and cooperation. *Public Administration Review*, 67(2). 290–301.

Carrington, K. (2005). Street-level discretion: Is there a need for control? *Public Administration Quarterly*, 29(1/2). 139–161.

Caruson, K., & MacManus, S. (2006). Mandates and management challenges in the trenches: An intergovernmental perspective on homeland security. *Public Administration Review*, 66(4). 522–536.

Chisholm, D. (1989). *Coordination without hierarchy: Informal structures in multiorganizational systems*. Berkeley: University of California Press.

deLeon, P., & deLeon, L. (2002). What ever happened to policy implementation? An alternative approach. *Journal of Public Administration Research and Theory*, 12(4). 467–492.

deLeon, P. & Varda, D. (2009). Toward a theory of collaborative policy networks: Identifying structural tendencies. *Policy Studies Journal*, 37(1). 59–74.

Edwards, G., & Sharkansky, I. (1978). *The policy predicament.* San Francisco, CA: W. H. Freeman.

Elmore, R. (1985). Forward and backward mapping: Reversible logic in the analysis of public policy. In K. Hanf & T. Toonen (Eds.), *Policy implementation in federal and unitary systems* (pp. 33–70). Boston, MA: Nijhoff Publishers.

Emerson, K., Nabatchi, T., & Balogh, S. (2012). An integrative framework for collaborative governance. *Journal of Public Administration Research and Theory*, 22(1). 1–29.

Fagan, P. (1997). Collaboration toward a more integrated national ocean policy: Assessment of several U.S. federal interagency coordination groups. *Dissertations & Thesis Full Text*, 58(11), 5874. (UMI No. 9815731)

Goggin, M. (1986). The "too few cases/too many variables" problem in implementation research. *The Western Political Quarterly*, 39(2). 328–347.

Goggin, M., Bowman, A., Lester, J., & O'Toole, L. (1990). *Implementation theory and practice: Toward a third generation.* Glenview, IL: Scott, Foresman/Little, Brown.

Goldman, H., & Intriligator, B. (1990). *Factors that enhance collaboration among education, health and social service agencies.* Paper presented at the Annual Meeting of the American Educational Research Association, Boston, MA (ERIC Document Reproduction Service No. ED 318 109).

Goldsmith, S., & Eggers, W. (2004). *Governing by network: The new shape of the public sector.* Washington, DC: Brookings Institution Press.

Gray, B. (1985). Conditions facilitating interorganizational collaboration. *Human Relations*, 38(10). 911–936.

Gray, B. (1989). *Collaborating: Finding common ground for multiparty problems.* San Francisco, CA: Jossey-Bass Publishers.

Hall, T., & O'Toole, L. (2000). Structures for policy implementation: An analysis of national legislation 1965–1966 and 1993–1994. *Administration and Society*, 31(6). 667–686.

Hall, T., & O'Toole, L. (2004). Shaping formal networks through the regulatory process. *Administration and Society*, 36(2). 186–207.

Hardy, C., Lawrence, T. B., & Grant, D. (2005). Discourse and collaboration: The role of conversations and collective identity. *Academy of Management Review*, 30(1). 58–77.

Hjern, B. (1982). Implementation research—The link gone missing. *Journal of Public Policy*, 2(3). 301–308.

Hjern, B., & Porter, D. (1981). Implementation structures: A new unit of administrative analysis. *Organization Studies*, 2(3). 211–227.

Howlett, M. (2003). *Studying public policy: Policy cycles and subsystems.* New York: Prentice Hall.

Huxham, C. (2003). Theorizing collaboration practice. *Public Management Review*, 5(3). 401–423.

Imperial, M. (2001). Collaboration as an implementation strategy: An assessment of six watershed management programs. *Dissertations & Thesis Full Text*, 62(02), 767. (UMI No. 3005481)

Imperial, M. (2005). Using collaboration as a governance strategy: Lessons from six watershed management programs. *Administration & Society*, 37(3). 281–320.

Intriligator, B. A. (1992). *Establishing inter-organizational structures that facilitate successful school partnerships.* Paper presented at the Annual Meeting of the American Education Research Association Boston, MA, April 16–20, 1990 (ERIC Document Reproduction Service No. ED 347 692).

Intriligator, B. A. (1994). Coordinating services for children and families: The organizational perspective. In R. Levin (Ed.), *Greater than the sum: Professionals in a comprehensive services model* (pp. 19–44). Washington, DC: ERIC Clearinghouse on Teacher Education.

Jennings, E. (1994). Building bridges in the intergovernmental arena: Coordinating employment and training programs in the American states. *Public Administration Review*, 54(1). 52–60.

Jennings, E., & Ewalt, J. (1998). Interorganizational coordination, administrative consolidation, and policy performance. *Public Administration Review*, 58(5). 417–428.

Jennings, E., & Krane, D. (1994). Coordination and welfare reform: The quest for the philosopher's stone. *Public Administration Review*, 54(4). 341–348.

Kapucu, N. (2006). Interagency communication networks during emergencies: Boundary spanners in multiagency coordination. *American Review of Public Administration*, 36(2). 207–225.

Keast, R., Brown, K., & Mandell, M. (2007). Getting the right mix: Unpacking integration meanings and strategies. *International Public Management Journal*, 10(1). 9–33.

Keast, R., Mandell, M., Brown, K., & Woolcock, G. (2004). Network structures: Working differently and changing expectations. *Public Administration Review*, 64(3). 363–371.

Kettl, D. (2003). Contingent coordination: Practical and theoretical puzzles for homeland security. *American Review of Public Administration*, 33(3). 253–277.

Kuska, G. (2005). Collaboration toward a more integrated national ocean policy: Assessment of several U.S. federal interagency coordination groups. *Dissertations & Thesis Full Text*, 66(12). (UMI No. 3200549)

Lambright, H. (1997). The rise and fall of interagency cooperation: The U.S. Global Change Research Program. *Public Administration Review*, 57(1). 36–44.

LaRocco, D. (1997). An analysis of the collaborative nature of state and local Part H interagency efforts and the consequent relationships between these levels of Part H governance. *Dissertations & Thesis Full Text*, 58(04), 1245. (UMI No. 9728849)

Linder, S., & Peters, B. (1987). A design perspective on policy implementation: The fallacies of misplaced prescription. *Policy Studies Review*, 6(3). 459–475.

Lipsky, M. (1980). *Street-level bureaucracy: Dilemmas of the individual in public services.* New York: Russell Sage Foundation.

Long, E. and Franklin, A. (2004). The paradox of implementing the Government Performance and Results Act: Top-down direction for bottom-up implementation. *Public Administration Review*, 64(3). 309–319.

Lundin, M. (2007). Explaining cooperation: How resource interdependence, goal congruence, and trust affect joint actions in policy implementation. *Journal of Public Administration Research and Theory*, 17(4). 651–672.

Mandell, M. (1994). Managing interdependencies through program structures: A revised paradigm. *American Review of Public Administration*, 24(1), 99–121.

Mandell, M. (1999). The impact of collaborative efforts: Changing the face of public policy through networks and network structures. *Policy Studies Review*, 16(1). 4–17.

Mandell, M., & Steelman, T. (2003). Understanding what can be accomplished through interorganizational innovations: The importance of typologies, context, and management strategies. *Public Management Review*, 5(2). 197–224.

Matland, R. (1990). *Reconciling top-down and bottom-up models of implementation.* Paper for Southern Political Science Association Meetings, Atlanta, Germany.

Matland, R. (1995). Synthesizing the implementation literature: The ambiguity-conflict model of policy implementation. *Journal of Public Administration Research and Theory*, 5(2). 145–174.

Mattessich, P., Murray-Close, M., & Monsey, B. (2001). *Collaboration: What makes it work?* Saint Paul, MN: Amherst H. Wilder Foundation.

May, P. (1995). Can cooperation be mandated? Implementing intergovernmental environmental management in New South Wales and New Zealand. *Publius*, 25(1). 89–113.

Maynard-Moody, S., & Musheno, M. (2000). State agent or citizen agent: Two narratives of discretion. *Journal of Public Administration Research and Theory*, 10(2). 329–358.

Mazmanian, D., & Sabatier, P. (1989). *Implementation and public policy.* New York: University of California.

McFarlane, D. (1989). Testing the statutory coherence hypothesis: The implementation of federal family planning policy in the states. *Administration and Society*, 20(4). 395–422.

McGuire, M. (2006). Collaborative public management: Assessing what we know and how we know it. *Public Administration Review*, 66. 33–43.

McNamara, M. W. (2008). Exploring interactions during multiorganizational policy implementation: A case study of the Virginia Coastal Zone

Management Program. *Dissertations and Thesis Full Text*, 69(11). (UMI No. 3338107)

McNamara, M. W. (2012). Struggling to untangle the web of cooperation, coordination, and collaboration: A framework for public managers. *International Journal of Public Administration*, 35(6). 389–401.

McNamara, M. W. (2016a). Collaborative management and leadership: A skill set for the entrepreneur. In J. C. Morris & K. Miller-Stevens (Eds.), *Advancing collaboration theory: Models, typologies, and evidence* (p. 116–132). London: Routledge Publishing.

McNamara, M. W. (2016b). Unraveling the characteristics of mandated collaboration. In J. C. Morris & K. Miller-Stevens (Eds.), *Advancing collaboration theory: Models, typologies, and evidence* (pp. 65–86). London: Routledge Publishing.

McNamara, M. W., Leavitt, W., & Morris, J. (2008). *Multiple-sector partnerships and the engagement of citizens in social marketing campaigns.* Paper for Annual Conference of the American Society for Public Administration, Dallas, TX.

McNamara, M. W., Miller-Stevens, K., & Morris, J. (2019). Exploring the determinants of collaboration failure. *International Journal of Public Administration*, 43(1). 49–59.

McNamara, M. W. & Morris, J. (2012). More than a "one-trick pony": Exploring the contours of a multi-sector convener. *Journal for Nonprofit Management*, 15(1). 84–103.

McNamara, M. W., Morris, J. C., & Mayer, M. (2014). Expanding the universe of multiorganizational arrangements: Contingent coordination and the Deepwater Horizon transportation challenges. *Policy & Politics*, 42(3). 345–367.

Meier, K., & O'Toole, L. (2003). Public management and educational performance: The impact of managerial networking. *Public Administration Review*, 63(6). 689–699.

Menzel, D. (1987). An interorganizational approach to policy implementation. *Public Administration Quarterly*, 11(1). 3–16.

Montjoy, R., & O'Toole, L. (1979). Toward a theory of policy implementation. *Public Administration Review*, 39(5). 465–476.

Morris, J. C., & Burns, M. (1997). Rethinking the interorganizational environments of public organizations. *Southern Political Review*, 25(1). 3–25.

Morris, J. C., Gibson, W. A., Leavitt, W. M., & Jones, S. C. (2013). *The case for grassroots collaboration: Social capital and ecosystem restoration at the local level.* Lanham, MD: Lexington Press.

Morris, J. C., McNamara, M. W., & Belcher, A. (2019). Building resilience through collaboration between grassroots citizen groups and governments: A case study. *Public Works Management & Policy*, 24(1). 50–62.

Nylen, U. (2007). Interagency collaboration in human services: Impact of formalization and intensity on effectiveness. *Public Administration*, 85(1). 143–166.

O'Leary, R., & Bingham, L. (2007a). Introduction. *International Public Management Journal*, 10(1). 3–7.

O'Leary, R., & Bingham, L. (2007b). Conclusion: Conflict and collaboration in networks. *International Public Management Journal*, 10(1). 103–109.

O'Leary, R., Choi, Y., & Gerard, C. (2012). The skill set of the successful collaborator. *Public Administration Review*, 72(SI). 570–583.

Olson, L. (1996). Discovering the dimensions of collaboration: An investigation of the reliability and validity of the modified Interagency Arrangement Model. *ProQuest Dissertations & Theses Full Text*, 63(05), 1638. (UMI No. 9709983)

Ospina, S., & Yaroni, A. (2003). Understanding cooperative behavior in labor management cooperation: A theory-building exercise. *Public Administration Review*, 63(4). 455–471.

O'Toole, L. (1983). Interorganizational cooperation and the implementation of labour market training policies: Sweden and the Federal Republic of Germany. *Organization Studies*, 4(2). 129–150.

O'Toole, L. (1986). Policy recommendations for multi-actor implementation: An assessment of the field. *Journal of Public Policy*, 6(2). 181–210.

O'Toole, L. (1989). Alternative mechanisms for multiorganizational implementation: The case of wastewater management. *Administration and Society*, 21(3). 313–339.

O'Toole, L. (1991). *Multiorganizational policy implementation: Some limitations and possibilities for rational choice contributions.* Paper for Workshop on Games in Hierarchies and Networks, Koln, Germany.

O'Toole, L. (1993). Interorganizational policy studies: Lessons drawn from implementation research. *Journal of Public Administration Research and Theory*, 3(2). 232–251.

O'Toole, L. (1995). Rational choice and policy implementation: Implications for interorganizational network management. *American Review of Public Administration*, 25(1). 43–57.

O'Toole, L. (1997). Treating networks seriously: Practical and research-based agendas in public administration. *Public Administration Review*, 57(1). 45–52.

O'Toole, L. (2000). Research on policy implementation: Assessment and prospects. *Journal of Public Administration Research and Theory*, 10(2). 263–288.

O'Toole, L., & Montjoy, R. (1984). Interorganizational policy implementation: A theoretical perspective. *Public Administration Review*, 44(6). 491–503.

Peters, G. (1998). Managing horizontal government: The politics of coordination. *Public Administration*, 76. 295–311.

Pressman, J., & Wildavsky, A. (1973). *Implementation* (3rd ed.). Berkeley: University of California Press.

Raelin, J. A. (1980). A mandated basis of interorganizational relations: The legal-political network. *Human Relations*, 33(1). 57–68.

Raelin, J. A. (1982). A policy output model of interorganizational relations. *Organization Studies*, 3(3). 2243–2267.

Reilly, T. (2001). Collaboration in action: An uncertain process. *Administration in Social Work*, 25(2). 53–74.

Robinson, S. (2006). A decade of treating networks seriously. *The Policy Studies Journal*, 34(4). 589–598.

Sabatier, P. (1986). Top-down and bottom-up approaches to implementation research: A critical analysis and suggested synthesis. *Journal of Public Policy*, 6(1). 21–48.

Saetren, H. (2005). Facts and myths about research on public policy implementation: Out-of-fashion, allegedly dead, but still very much alive and relevant. *Policy Studies Journal*, 33(4). 559.

Schlossberg, M. (2004). Coordination as a strategy for serving the transportation disadvantaged: A comparative framework of local and state roles. *Public Works Management and Policy*, 9(2). 132–144.

Schofield, J. (2001). Time for a revival? Public policy implementation: A review of the literature and an agenda for future research. *International Journal of Management Reviews*, 3(3). 245–263.

Stoker, R. (1991). *Reluctant partners*. Pittsburgh, PA: University of Pittsburgh Press.

Thatcher, C. (2007). A study of interorganizational arrangement among three regional campuses of a large land-grant university. *Dissertations & Thesis Full Text*, 68(03). (UMI No. 3255178)

Thomas, C. (1997). Public management as interagency cooperation: Testing epistemic community theory at the domestic level. *Journal of Public Administration Research and Theory*, 7(2). 221–246.

Thompson, J. (1967). *Organizations in action: Social science bases of administrative theory*. New Brunswick, NJ: Transaction Publishers.

Thomson, A., & Perry, J. (2006). Collaboration processes: Inside the black box. *Public Administration Review*, 55. 20–32.

Van de Ven, A., Delbecq, A., & Koenig, R. (1976). Determinants of coordination modes within organizations. *American Sociological Review*, 41(2). 322–338.

Van de Ven, A., & Walker, G. (1984). The dynamics of interorganizational coordination. *Administration Science Quarterly*, 29(4). 598–621.

Van Horn, C. (1979). *Policy implementation in the federal system*. Lexington, KY: Lexington Books.

Wise, C. (2006). Organizing for homeland security after Katrina: Is adaptive management what's missing? *Public Administration Review*, 66(3). 302–318.

Wise, C., & Nader, R. (2002). Organizing the federal system for homeland security: Problems, issues, and dilemmas. *Public Administration Review*, 62. 44–57.

Wood, D., & Gray, B. (1991). Toward a comprehensive theory of collaboration. *Journal of Applied Behavioral Science*, 27(2). 139–162.

3   An Empirical Test of the Multiorganizational Interaction Model

This chapter focuses on gaining conceptual clarity by operationalizing elements that may be used to distinguish between cooperation, coordination, and collaboration. Our intent is that this conceptual foundation may provide a starting point for continued exploration into better understanding these terms. The strength of this foundation is based on a combination of elements identified to allow interaction terms to be compared empirically. Clarity of elements within the framework provides a needed structure for distinguishing between different types of interaction and enhances its potential for application in various settings.

A Note about Study Methods

Data are collected through semistructured interviews and a review of documents. In total, 34 interview transcriptions and eight documents are reviewed to explore interactions during multiorganizational policy implementation. Textual data from these interviews and documents are analyzed using a predetermined coding scheme aligned with the operationalizations of the Multiorganizational Interaction Model (MIM). Upon completion of coding, we organize the data based on each element of the model. Results of the analysis are organized by the study's research questions and are presented throughout this chapter.

We developed a multi-pronged qualitative methodology to collect and analyze data for the empirical portion of this study. We employed semistructured interviews to gather in-depth information from administrators representing a variety of organizations within the network implementing the program. A pre-structured coding scheme, based on sensitizing themes, helped link the textual data collected to operationalizations of the MIM. The interview protocol included open-ended questions that generally speak to the four constructs of the MIM. These questions were written broadly so as not to bias participant responses or lead them to provide answers that align with the model's operationalizations. Probing questions were used to seek additional information or clarify previous responses. While these questions were also open-ended, they were phrased to guide inquiry more specifically to the operationalizations of the model's constructs.

The sampling frame consisted of participants representing the organizations involved in implementing the program. Interview participants were selected using snowball sampling. The interviews were recorded electronically to allow the interviewer to engage conversationally with the interviewee and to probe for clarification when needed; the use of audio recordings also ensured accuracy of data. The recordings were transcribed verbatim, and the transcriptions emailed to interviewees to provide them an opportunity to make revisions to the document. This technique helped ensure authenticity and accuracy of the findings. We also reviewed a number of documents related to program design, implementation, and operation. Analysis of these documents allowed for triangulation of the themes developed from the interviews. They were also useful for exploring the context in which interactions occur, the history of interactions between organizations, and the formality in which interactions were initiated.

We reviewed all textual data, focusing our analysis on patterns that align with categories of the pre-structured coding scheme. Operationalizations of the MIM's constructs provided the basis for the coding scheme. Data analysis initially occurred through a deductive approach in which an existing framework was used to verify theory and assess linkages with qualitative data. Coding labels were used to identify each variable in the model and each type of interaction. These shorthand designations were used to assign meaning to data and guide analysis by linking data with the study's research questions.

Once the textual data were coded, the data were numerically aggregated or left in textual form. For numerical aggregation, we used quantification to identify the number of times a particular phenomenon emerged within the content of the interviews or organizational documents. As patterns emerged, the data were reduced and numerically aggregated into the interaction categories of cooperation, coordination, or collaboration. We also reviewed the transcripts in textual form; comments from interviewees provided context to the research that would otherwise have been lost. Emerging patterns from this textual data were also mapped to the categories within the predetermined coding scheme to enrich numerical aggregation. We employed content analysis to identify emerging data patterns that did not fit into the predetermined categories of the coding scheme. Corroboration of complementary data sources enhanced the authenticity of this study's findings. Triangulation of data sources was established by comparing data gathered from documents with data gathered from interviews. Data collected from both sources were compared to address inconsistencies, corroborate findings, and illuminate different approaches to the same phenomena.

Participants for this study were selected using a snowball sampling strategy. Interviewees represent 15 organizations across the public sector (federal, state, and local) and nongovernmental organizations. A breakdown of participants based on organizational sector affiliation is presented in Table 3.1. More specific identification of personnel is not revealed to preserve anonymity.

Table 3.1 Network of Organizations in the Virginia Seaside Heritage Program

Organizational Type (# of Participants Interviewed)	*Specific Organizations in the Network*
Federal agencies (1)	U.S. Fish and Wildlife Service—Eastern Shore National Wildlife Refuge
State agencies/programs (18)	Secretariat of Natural Resources Department of Environmental Quality Coastal Zone Management Program Marine Resources Commission Habitat Management Oyster Conservation Department of Conservation and Recreation Planning and Recreation Division Natural Heritage Division Department of Game and Inland Fisheries Watchable Wildlife Division Nongame Division
Local government (6)	Accomack County Planning Soil and Water Conservation District Northampton County Accomack-Northampton Planning District Commission

Organizational Type (# of Participants Interviewed)	Specific Organizations in the Network
Nongovernmental organizations (9)	The Nature Conservancy
	Eastern Shorekeeper
	Southeast Expeditions
	Cherrystone Aquafarms
	College of William & Mary[a]
	Institute of Marine Science[a]
	Center for Conservation Biology[a]
	University of Virginia[a]

a Although these academic institutions are state sponsored, they operate autonomously as individual organizations.

Analysis of the Multiorganizational Interaction Model

Does the Multiorganizational Interaction Model (MIM) help explain interactions in a policy implementation setting? We employ content analysis to examine the data through the lens of the theoretical model. Tables presented in this section are based on the combined data collected through interviews and document review. Data are organized into categories of the predetermined coding scheme aligned with the operationalizations of the theoretical model.

The utility of the MIM to help explain interactions in a policy implementation setting is explored by identifying patterns within the textual data that speak to the following four constructs: interorganizational policy objective, interorganizational infrastructure, interorganizational procedures, and organizational management. Elements within all four constructs are present in the coding of interviews and documents. Congruence between the textual data gathered and the operationalizations of the model's constructs are of particular interest. Pattern matching is used to determine the extent to which the data's empirical patterns match the model's theoretical patterns.

Analysis of the Interorganizational Policy Objective Construct

For the Virginia Seaside Heritage Program (VSHP), the interorganizational policy objective represents the program goals the organizations are working together to achieve. More specifically, the program goals focus on the preservation and management of natural resources on Virginia's Eastern Shore. In the MIM, the interorganizational policy objective construct is characterized by four variables: time, difficulty, role of single organization, and the impetus for collective action. Data, gathered through interviews and documents, are coded based on the operationalizations of each variable. Twenty-nine percent of the elements identified in this study are within the policy objective construct. In terms of the utility of the interorganizational policy objective construct within the MIM, the data gathered through this study suggests that the time, difficulty, role of single organizations, and impetus for collective action variables help to explain interactions between organizations. Of these four variables, impetus for collective action accounts for 60% of the elements mentioned during interviews or within organizational documents. In the VSHP, the ways in which the multiorganizational arrangement develops is mentioned more than four times as often as the other variables within this construct.

There seems to be two driving forces for the development of the multiorganizational arrangement within this study. First, organizations work together because it helps them achieve their individual or mutual goals. These two elements account for 36% of the impetus for the collective action variable. Discussion during an interview reveals the importance placed on these elements. "It behooves us to work with the Coastal Zone Management Program because the whole is greater than the sum of the parts. So working together helps us achieve what we need to do as an organization."

Second, organizations work together based on the presence of a legislative mandate/grant contract or a convener. These two elements account for another 42% of the impetus for collective action variable. The Virginia Coastal Zone Management (VCZM) Program convenes the VSHP, and its staff plays a particularly important role in this study. One interviewee suggested that the VCZM Program staff "d[oes] a good job of bringing the right people in and helping them understand that creating this regional coalition was not only possible but beneficial to everyone." In addition, the presence of a stable funding stream helps lure organizations to the table. The VCZM Program distributes about $500,000 annually to organizations within the program. An interviewee suggests that the VCZM Program draws attention from organizations because of the grant money available. The money that the VCZM Program brings to the table enables personnel to make their projects a reality without having to spend time "chasing funding." Other interviewees agree by saying, "Longevity of funding is critical," and "The money enables the work to move forward." Another interviewee provides a reason for the criticality of funding, "Virginia spends less than 1% of its budget on natural resources. The environment is the underdog in the state budget and in the national budget when it comes down to it." In conjunction with the presence of a convener, the benefits associated with a source of funding should not be underestimated.

There are elements associated with this construct that are not mentioned or are only mentioned one time throughout interviews and documents. In terms of the time variable, the "short-term" and "longer-term" elements are only mentioned one time or not at all respectively. Absence of these elements may be attributed to the inherently long time frame associated with environmental work. In terms of the difficulty variable, the "simple task" element is only mentioned one time. Although the literature suggests that organizations work together to pursue relatively simple tasks (see, for example, Keast, Brown, & Mandell, 2007; O'Leary & Bingham, 2007), it is not the case for organizations in this study. While almost a quarter of interview participants indicate that some tasks could be considered to be relatively simple, their responses are framed in a larger context indicating that the number of people involved in the program, tremendous diversity of tasks associated with the goals of the program, and interdependencies between tasks make the situation increasingly more complex. These elements may occur more frequently in a different policy arena and should be further explored in other settings before they are removed from the model.

Another element that is seemingly not found in the coding of the interviews and documents is "changes in external factors trigger organizations to search for new solutions." Changes in external factors such as rising sea level, increases in population, and pressures to develop land are discussed. However, these situations are discussed in the sense that they are so complicated that organizations have to work together because no organization can address these problems individually (Gray, 1985). Therefore, these references are coded under the element of "situations of crisis" or "cannot achieve the desired goal without working together." Changes in external factors seem to generate collaborative arrangements because organizations cannot achieve the desired goal without working together. The findings from this study do not align with the literature, which suggests that cooperative arrangements may be triggered by external factors and the desire to avoid negative impacts associated with these factors (see, for example, Ospina & Yaroni, 2003). As a result of this research, the "changes in external factors" element is removed from the impetus for collective action variable under cooperation and acknowledged as a supporting statement for the existing elements within the same variable in collaboration.

Analysis of the Interorganizational Infrastructure Construct

The interorganizational infrastructure construct focuses on the ways in which relationships within the multiorganizational arrangement are generated and structured. Five variables characterize the construct of interorganizational infrastructure: design, formality of the agreement, organizational autonomy, policy authority, and key personnel. Data gathered through interviews and document review are coded based on the operationalizations of each variable. Nineteen percent of the elements identified in this study are within the interorganizational infrastructure construct.

In terms of the utility of the interorganizational infrastructure construct within the MIM, data gathered through this study suggests that the design, formality of the agreement, organizational autonomy, policy authority, and key personnel variables help to explain interactions between organizations. Of these four variables, the elements pertaining to the formality of the agreement variable are most emphasized as it accounts for 32% of the elements within the infrastructure construct. In the VSHP, it appears that there are two prominent ways in which organizations agree on their roles and responsibilities within the arrangement. First, the "contracts or nonfinancial agreements formalize relationships" element accounts for more than one-third of the number of times elements within the formality of the agreement variable are mentioned. In their research on the Job Opportunities and Basic Skills Program, Jennings and Krane (1994) find that contractual or nonfinancial agreements are used to generate the relationships between organizations. Their finding is consistent with this study. Within the VSHP, grant contracts are also used to establish and sustain relationships between organizations.

For example, the Accomack-Northampton Planning District Commission (PDC) receives a technical assistance grant from the VCZM Program annually. This grant requires a minimum standard of interaction between the PDC and the local governments.

An interviewee describes the importance placed on grant contracts in formalizing relationships: "The grant contract provides the conduit for the flow of information from the state through the planning district commission to the localities. And just as importantly, from the localities back up to the state." In addition, an organization may receive grant funding from the VCZM Program and subcontract with another organization to complete a particular project. This can be seen in the construction of an observation platform in the town of Willis Wharf. The Virginia Department of Game and Inland Fisheries (DGIF) received grant funding from the VCZM Program to design and build the observation platform. While DGIF completed the design of the platform, they subcontracted with Northampton County to construct it. In this instance, the original grant and subsequent contracts define each organization's role and responsibilities. Much like the research conducted by Jennings and Krane (1994), contractual arrangements between agencies outline roles and responsibilities. In this study, over 40% of interviewees mention the use of grant contracts or nonfinancial agreements to formalize relationships.

Second, organizations informally work together to achieve individual goals. This element accounts for another quarter of the times elements within this variable are mentioned. Informal interactions are prevalent among the participants involved in the VSHP, and relationships go well beyond the stipulations in grant contracts. There appear to be two explanations to support these informal interactions. First, there are long-standing relationships between this group of people, and they work with each other on a variety of projects outside of the VSHP. An interviewee describes these long-standing relationships as follows: "There are a lot of people who have been on the Coastal Policy Team for the last 20 years. It's the continuity of the relationships that have been really helpful." This point is supported in another interview:

> So there [is] a core group that has gotten really good at working together over the last 15 years. When different things come up, we know to call each other... Having those long standing relationships really helps in terms of pulling the partners together.

Comments made by this interviewee convey the evolutionary nature of relationships in the implementation network. As the specifics of a situation emerge, organizations with expertise in the necessary areas are brought together. In addition, geographic proximity lends itself to informal interactions. An interviewee explains how opportunities for informal interactions are created among individuals involved in implementing the VSHP:

> There is a physical opportunity for people that live on Virginia's Eastern Shore. There are only two counties and there is really only one highway. And that helps. There are only 50,000 people on the Virginia portion of the Eastern Shore. So you see people at church and the grocery store.

Almost 50% of the participants in the VSHP live and work on the Eastern Shore. This proximity is an important factor to acknowledge within the multiorganizational arrangement.

There are elements associated with this construct that are not mentioned or are only mentioned one time throughout interviews and documents. In terms of design, the "interagency staff is unnecessary" element is not mentioned. One reason that this element may not be mentioned is because it seems closely related to the "organizations work within their existing organizational structures" element within the cooperative design. After additional review, it may be too difficult to distinguish these two elements. The literature suggests that cooperative interactions occur informally, and limited connections can be established through existing organizational structures (see, for example, Keast, Brown, & Mandell, 2007; Reilly, 2001). Since this type of interaction typically occurs between actors at the lower levels of organizations (Keast, Brown, & Mandell, 2007), it is implied that an interagency staff is not necessary. This study supports this assumption. As a result, the "interagency staff is unnecessary" element is removed from the design variable within the model.

In addition, the "centralization may involve program reorganization or consolidation" element within the design variable is not mentioned in this study. Formalized mechanisms, such as contractual or nonfinancial agreements, account for more than 10% of the elements present from this construct; however, centralization is rarely mentioned. The formalization within this program comes from grant contracts rather than a hierarchical design. Despite what is said in the literature (see, for example, Boston, 1992; Jennings, 1994), a hierarchical structure is not used in this study to enhance centralization based on a desire to reorganize or minimize duplication. Instead, the presence of a formalized structure is associated with the distribution of money. Since the absence of this element may also be attributed to the network design of the program, this element should be further explored in other settings before it is removed from the model.

The "organizations maintain individual authority over the policies that govern their respective organizations" element is only identified once in the interviews and documents. This may be due to the presence of the Coastal Policy Team (CPT), which is comprised of resource administrators and managers from each of the state agencies involved in the VSHP. As the literature suggests, the policies that govern individual organizations are maintained (Keast, Brown, & Mandell, 2007). Each agency operates under specific regulatory authorities and legal policies. However, this authority does not appear in the forefront of the data collected. Instead, interviewees place emphasis on the discretion given to the resource administrators and managers representing the CPT. Within the boundaries of their regulatory authorities, representatives make governing decisions to guide their divisions and the collective group. An interviewee explains the relationship between individual agency authorities and collective governing in the following manner:

> The CPT determines what the focal area is going to be and what is going to be done within that focal area. How particular projects are implemented comes under the regulatory functions of the agencies involved. But the program itself, the way it [i]s set up and the goals identified, that [i]s decided by the CPT.

This presence of discretion in collaborative interactions is supported by the literature (see, for example, Mattessich, Murray-Close, & Monsey, 2001). An interview participant explains how this discretion is used, "by creating a Coastal Policy Team, you do have upper level administrators involved, but they are the administrators focused on the resources issues. So we issue good judgment as to where the resources need to be applied."

Discretion allows managers to work across organizational boundaries to jointly develop policies. Despite the literature coupling hierarchical authority with bureaucratic organizations, implementation of the VSHP does not require significant time commitments from upper level personnel. Instead, resource administrators have discretion to make collective decisions with partners. Several interviewees attribute this presence of discretion to the success of the VSHP. Discretion may be given more freely when programs are successful. Since collaborative decisions require discretion, sustaining these interactions may also require a certain amount of success. Although the element of "organizations maintain individual authority over the policies that govern their respective organizations" is not prevalent in this study, it may be important in a different setting. This element remains in the model for further exploration.

Analysis of the Interorganizational Procedures Construct

The interorganizational procedures construct focuses on the processes developed to support operations or sustain relationships within the multiorganizational arrangement. Five variables characterize the construct of interorganizational procedures: information sharing, decision making, resolution of turf issues, resource allocation, and systems thinking. Thirty percent of the elements identified in this study are within the interorganizational procedures construct.

In terms of the utility of the interorganizational procedures construct within the MIM, data gathered through this study suggests that the information sharing, decision making, resolution of turf issues, resource allocation, and systems thinking variables help explain interactions between organizations. Of these four variables, elements pertaining to the information sharing and resource allocation variables are most prevalent. The information-sharing variable accounts for 34% of the elements within the interorganizational procedures construct. In the VSHP, there are three ways in which the multiorganizational arrangement develops processes to support operations or sustain relationships.

First, dialogue is maintained through informal relationships. This element accounts for almost one-third of the number of times elements are mentioned within the information-sharing variable. Its presence is consistent with the literature, which indicates that informal communication channels are used to maintain dialogue and share information among participants (see, for example, Ospina & Yaroni, 2003). Participants implementing the VSHP often communicate through channels such as email, ad hoc working groups, or by seeing each other in the field. An interviewee suggests that visibility between field workers is important for communication: "I'll see partners on the dock or on the water or in the coffee shop. And we talk about what is going on with different projects."

Second, participants frequently communicate to support operations or sustain relationships within the interorganizational arrangement. This element accounts for almost one-fifth of the number of times elements are mentioned within this variable. The presence of this element aligns with the literature, which emphasizes how open communication can reduce information asymmetries (see, for example, Thomson & Perry, 2006). An interview participant explains communication among partners in the following manner: "We are always talking to each other and bringing each other in on different projects... We come to the table on a regular basis. So it keeps that partnership and the relationships going."

Third, understanding between organizations is enhanced by a willingness to share information about organizations, which may include what can or cannot be offered to the collective group. This element accounts for more than one-fifth of the number of times elements are mentioned within this variable. Its presence within the study aligns with the literature and emphasizes the need for organizations to share information in order to create a base of common knowledge and promote understanding (see, for example, Keast, Brown, & Mandell, 2007; Imperial, 2001). An interviewee acknowledges a high level of understanding among organizations:

> We've gotten to be extremely good coworkers even though we are in different agencies. We know what each other can do, we know what talents we can bring to the table, we know the expertise that we each have, and we know when we can work together on what things.

In the interorganizational procedures construct, there are elements that are not mentioned or are only mentioned one time throughout interviews and document review. In terms of decision making, the "centralized" element is not mentioned. The absence of this element may be due to the way in which natural resource agencies are structured in Virginia. This structure is described by a participant: "In Virginia, we don't have a single resource agency. We have a number of agencies within a secretariat of natural resources but each has their own agency head, their own budgets, and their own specific missions." Virginia's natural resource agencies have equal representation on the CPT. Since distinct legal authorities guide each agency, no organization has the authority to tell another organization what to do. Therefore, decision making is not structured to be a centralized process. In addition, the very nature of environmental sustainability does not lend itself to centralized solutions. An interviewee suggests that nonregulatory environmental issues, such as sustainability, do not lend themselves to command and control decisions: "Other environmental issues like water quality monitoring, are much more command and control. It is the land-based, nonpoint source sustainable ecosystem stuff that does not have that sort of command and control." Since the absence of this element may be specific to this case study, it should be further explored before it is removed from the model.

In addition, the "resources allocated by balancing needs of group and individual organizations" element within the resource allocation variable is only mentioned one time in this study. When looking at the other elements within collaborative resource allocation, this element may not be mentioned for two reasons. First, participants may assume that they balance these needs when referring to the "pooled resources" element within this variable. As organizations determine what resources they can contribute to the collective unit, they likely balance their organization's needs with those of the larger group. Second, a high degree of alignment between organizational missions may make a balancing of needs unnecessary. An interviewee describes mission alignment in the following way: "Most of the projects we are working on are related directly to our agency's mission." In this study, each organization's interests align nicely with the program's broader goals of habitat conservation and restoration. Therefore, a balancing of needs is unnecessary because both sets of needs are simultaneously fulfilled. This element is combined with the "pooled resources" element within this variable in the final version of the MIM.

Analysis of the Organizational Management Construct

The organizational management construct focuses on the way in which behaviors within and between member organizations support the interorganizational arrangement. Five variables characterize the construct of organizational management: incentives, commitment, trust, risk taking, and willingness to change. Data, gathered through interviews and documents, are coded based on the operationalizations of each variable. Twenty-two percent of the elements identified in this study are within the organizational management construct.

In terms of the utility of the organizational management construct within the MIM, the data gathered through this study suggests that the incentives, commitment, trust, risk taking, and willingness to change variables help to explain interactions between organizations. Of these five variables, elements pertaining to the commitment variable are most emphasized within this construct; they account for 31% of the elements identified. In the VSHP, there are three prominent ways in which support for the interorganizational arrangement is generated within and between organizations.

First, a history of supportive behavior or long-standing relationships generates support for the interorganizational arrangement. This element accounts for almost 90% of the number of times elements are identified within the trust variable, and it accounts for more than 40% of the number of times elements are identified within the entire construct. In addition, this element is mentioned by more than 80% of participants during interviews. Findings are consistent with the literature, which indicates that a history of interactions between organizations enhances trust between organizations (see, for example, Mandell & Steelman, 2003). An interviewee describes the history between organizations involved in implementing the VSHP in the following manner:

> The secret of success [is] the continuity of the personnel over time. And that is not something that you can really control, that is just luck. And I think the fact that we have known each other for about 20 years now and we know what we are each about.

Second, funding provided through statutes or grant contracts generates support for the multiorganizational arrangement. In terms of the VSHP, funding incentives come from grant contracts. Although the literature identifies the potential for these incentives to occur through statutory provisions (see, for example, Jennings, 1994), that is not the case in this study. In fact, an interviewee specifically mentions that unfunded mandates often promulgate requirements pertaining to environmental sustainability. "The state gives the local governments the authority to do something, but they don't provide funding to support those initiatives." Due to the lack of funding from statutes, this element is revised to say "grant contracts may provide funding" in the final MIM.

Third, support for the multiorganizational arrangement is generated when the collective interest serves the individual interests of the organizations involved. This element accounts for more than one-fifth of the number of times elements are identified within this variable. Its presence within the study aligns with the literature, which recognizes that commitments to the collective arrangement do not diminish commitments to individual organizations (see, for example, Keast, Mandell, Brown, & Woolcock, 2004). Discussion during an interview indicates that the collective interest is attained through individual organizational interests: "We are meeting our agency's objectives but we are also furthering the whole effort." The findings from this study also align with the literature by acknowledging that the extent to which the collective interest serves the interests of individual organizations determines the extent to which they are willing to support collective endeavors (see, for example, Imperial, 2001; Thomson & Perry, 2006). Interactions within the collective group not only enable organizations to meet their objectives, but they also enable the collective group to surpass the level to which they would meet these objectives when working individually. An interviewee explains how multiorganizational interactions create opportunities for individual organizations:

> This program gives us the opportunity to do work that we otherwise would not be able to do and achieve a part of our mission that would otherwise not be possible. It is an opportunity to be successful in a way that would be impossible otherwise.

Within the organizational management construct, there are elements not mentioned or only mentioned one time throughout interviews and documents. In terms of the commitment variable, the "encouraged by supervisory administrative body" element is only mentioned one time. Jennings' (1994) research suggests that 70% of the administrators involved in the study indicate that leaders play an important role in encouraging coordination. In the VSHP, interviewees and documents do not suggest this same level of importance. The absence of this element may be attributed to the degree to which participants want to protect the resource and the program's success. An interview participant expresses the multiorganizational arrangement's commitment:

> All the people within these organizations care so deeply about the place—every one of them. I can't think of one person in th[e] VSHP partnership that I would say is not just deeply and personally committed to saving this place and making it better.

This finding suggests that if there is a high degree of personal commitment within organizations, then encouragement from a supervisory body may be less necessary.

In addition, the absence of involvement from top management may be attributed to the success of the program. An interviewee suggests that successful projects, coupled with zero complications, do not require great involvement from top leadership: "Trust is built through successful accomplishment of various projects that we work on and positive reinforcement." Discussion during another interview also reveals a relationship between program success and managerial trust. "People above us see and hear about the success of the program. And as long as they are hearing that, they aren't going to get involved. They'll say 'good job, keep doing it.'" Absence of this element could be specific to this study because individuals may commit more easily to environmental issues. Before this element is removed from the model, it is worth further exploration in a different setting.

In terms of the trust variable, the "leaders work closely to create relationships based on trust" element is only mentioned one time and the "trust relationships are not required but can develop" element is not mentioned at all. The irrelevance of these elements may be due to high levels of trust that permeate organizational boundaries. When personnel work together for great lengths of time, as they have in this study, leaders may not need to facilitate relationships based on trust. The literature's emphasis on organizational leaders developing trust amongst each other (see, for example, Jennings & Krane, 1994) is simply not found in this study. Since these results may be particular to the history of long-standing relationships between participants of the VSHP, these elements are left in the model for further study.

Within the risk-taking variable, six of the seven elements are only mentioned one time or not at all during interviews or document review. The only repeated element within this variable is "low levels of risk." As the literature suggests, partner organizations undoubtedly generate dependencies as integrated policies and operations form (see, for example, Imperial, 2005). However, this study does not support the assertion that collaborative arrangements involve high levels of risk (see, for example, Keast et al., 2004). In fact, not one interviewee suggests that there are high levels of risk associated with working together. On the contrary, an interviewee emphasizes low levels of risk within the multiorganizational arrangement: "The risk is fairly low. And the trust is fairly high." This sentiment is echoed throughout many interviews. The literature fails to acknowledge that the risk associated with dependencies can be minimized through high levels of trust and long-standing relationships. The emphasis that this study places on low levels of risk mitigates the presence of all other elements within this variable.

Regardless of the type of interaction, we suggest that people will work together if it benefits their own organization to some degree. Findings from this study suggest that low levels of risk are associated with each type of interaction. If the interaction is too risky, the organization is not likely to become involved. Although cooperative interactions are typically associated with not engaging in risk-taking behavior (see, for example, Keast, Brown, & Mandell, 2007), higher levels of risk associated with coordinative and collaborative interactions may be mitigated by other factors. Therefore, the risk-taking variable is eliminated from the final version of the MIM.

Reviewing the MIM

While ambiguities within the original model make it difficult to distinguish between the elements associated with the three types of interactions, the findings from this study eliminate ambiguities while furthering interorganizational theory in two ways: (1) the MIM helps explain interactions between organizations in this policy implementation setting; and (2) a finalized version of the MIM is presented. Data from this study suggests that the MIM is helpful in explaining interactions in multiorganizational arrangements. After a review of the public administration and interorganizational theory literatures, we identified 107 elements to operationalize 19 variables within the four constructs of the MIM. Of these elements, only eight are not identified in this study's interview transcriptions and documents. Therefore, more than 92% of the elements put forth within the revised model are identified in this study.

Of the eight elements not identified, the following suggestions are made throughout this chapter to change four of the elements in the model: (1) we remove the "change in external factors" element from the impetus for collective action variable under cooperation and acknowledged as a supporting statement for collaborative elements within the same variable; (2) the "interagency staff is unnecessary" element is removed from the model because it is too difficult to decipher from the "organizations work within their existing organizational structures" element; (3) the "depart from normal behavior" element is removed from the model because the risk-taking variable is eliminated; and (4) the "high levels of risk" element is removed with the elimination of the risk-taking variable. While these elements may be potentially relevant in other settings, the data gathered from this study strongly suggests their removal from the model. The following four elements are left in the model for further research: (1) the "longer-term" element within the time variable; (2) the "centralization may involve program reorganization or consolidation" element within the design variable; (3) the "centralized" element within the decision-making variable; and (4) the "trust relationships are not required but can develop" element within the trust variable.

In applying the policy implementation and interorganizational theory literatures to revise the model's operationalizations, clear distinctions between the three interaction terms significantly improve the theoretical model. The findings suggest that the MIM can be used to enhance theoretical consistency and improve communication within the interorganizational theory literature. An example of how this model enhances theoretical consistency can be seen within the coordination literature. Terms such as "mobilization coordination" (Van de Ven & Walker, 1984), "contingent coordination" (Kettl, 2003), and "intermittent coordination" (Mandell & Steelman, 2003) are used to describe less formal coordination in which ad hoc relationships evolve based on the specifics of a given project. Throughout interviews from this study, elements associated with these descriptions never once align within the model's operationalizations of coordination. Instead, informal and ad hoc relationships are better represented by the operationalizations of cooperation while the evolutionary nature of the network is better represented by the operationalizations of collaboration. This suggests that the ways in which researchers identify informal views of coordination are actually not coordination at all. Instead, a different form of interaction better explains these views. In using the interorganizational theory literature to clearly distinguish between each type of interaction within the model, the application of the MIM in turn fosters theoretical consistency within the literature.

Findings from this analysis also enhance the authenticity of the MIM. In this study, we do not identify any patterns that do not fit into the categories of the predetermined coding scheme. Interviewees were given an opportunity to identify other factors by asking them the following question: "Are there any other factors that would help me understand the interactions between organizations when implementing the VSHP?" Typically, participants used this opportunity to reiterate a point previously made during the interview. In no instance did a participant mention factors that are not already captured within the theoretical model. The final draft of the MIM is presented in Tables 3.2–3.5.

Table 3.2 Final Variable Operationalizations: Interorganizational Policy Objective Construct

Variable	Cooperation	Coordination	Collaboration
Time	Short-term	Longer-term	Long-term, evolutionary nature
Difficulty	Simple task	Multifaceted tasks, repeatable	Complex tasks that are highly varied and diverse; or situations of crisis
Role of single organizations	Organizations are independent; it is possible for them to accomplish the task individually.	Organizations require some assistance from other organizations to accomplish individual goals/missions.	Organizations are interdependent; each organization is one element of the larger system.
Impetus for collective action	Typically voluntary, organizations initiate collective action because it is helpful to their world of work and it builds capacity that serves the individual organization.	Voluntary or mandated, linkages are mobilized because compatible mission areas mutually increase abilities to achieve individual goals. An interagency liaison or boundary spanner may forge these relationships to meet resource needs or shared interests. Legislative mandate or grant contracts may enhance cohesion or minimize duplication.	Voluntary or mandated, organizations with mutual or complementary interests come together because they cannot achieve the desired goal or address the identified problem without working together. Organizations share responsibility for tasks that are interconnected or cannot be accomplished individually. A lead agency or convening organization brings relevant stakeholders together and legitimizes collective action.

Table 3.3 Final Variable Operationalizations: Interorganizational Infrastructure Construct

Variable	Cooperation	Coordination	Collaboration
Design	Individuals work independently within existing organizational structures.	Each organization's hierarchical structure is used to centrally manage specialized roles and responsibilities. Centralization may involve reorganization or consolidation of programs/activities.	Partner organizations jointly develop shared power arrangements to support mutually beneficial interests. New program structures are developed based on the needs of a specific policy/goal. An administrative staff element is present to sustain collective efforts.
Formality of the agreement	Individual organizations informally agree to work together to achieve individual goals.	Mechanisms, such as contractual or nonfinancial agreements, formalize relationships between organizations. Agreements, clearly identifying each organization's roles and responsibilities, are often developed and/or reviewed by a higher authority.	Key stakeholders jointly draft a shared purpose and develop a course of action based on mutually agreed upon roles and responsibilities, rules, goals, and organizational boundaries.
Organizational autonomy	Organizations are fully autonomous.	Organizations are semiautonomous; individual organizations require some assistance from other organizations to achieve goals.	Organizations are not autonomous; operations within organizations are intertwined.
Policy authority	No interorganizational policy decisions are made. Preexisting policies, established by the individual organizations, are followed.	Organizations maintain individual authority over the policies that govern their respective organizations. Policies pertaining to coordinated efforts may be developed, but they are compatible with the policies already established within the individual organizations.	Partner organizations jointly develop policies and procedures that govern the collective group. Interorganizational policies and procedures include working rules that specify which stakeholders can make decisions, who will guide collective actions, and the distribution of costs/benefits.
Key personnel	Organizational leadership is not involved in decisions to work together.	There is a distinction between leaders and managers; leaders within each organization make decisions while managers implement and administer these decisions. A facilitator may be identified to coordinate actions at the local level.	Although no one is typically in charge, a lead organization may propose policies/rules to which the collective group must mutually agree to implement. Membership, role definitions, and responsibilities adapt to the task at hand. Each role is considered equally important.

Table 3.4 Final Variable Operationalizations: Interorganizational Procedures Construct

Variable	Cooperation	Coordination	Collaboration
Information sharing	Dialogue is maintained through informal relationships between participants. Basic information is initially shared. Continuous dialogue creates opportunities for discussing a wider range of topics.	Formal and informal communication channels are used to link vertical and horizontal organizational levels.	Emphasize open and frequent communications between partners to reduce information asymmetries. Formal and informal channels are used to widely disseminate information concerning the collective group. Understanding enhanced by a willingness to share information about individual organizations and what can/cannot be offered to the collective group.
Decision making	Decisions are made independently; rules that guide collective decision making are not necessary.	Centralized decision making is practiced; a lead organization(s) dominates the decision-making process.	Participative decision making based on consensus and compromise; generates rules to govern activities and relationships between organizations. Representatives have latitude to negotiate rules and deliberate agreements to identify common ground.
Resolution of turf issues	Turf issues between participating organizations are avoided based on organizational tendencies to function independently.	A neutral facilitator, outside convener, or full-time coordinator is employed to resolve turf issues.	Conflicting roles based on incongruent demands from individual organization and group. Consider adjusting policies and procedures to reduce conflict while maximizing common ground.
Resource allocation	Discretionary funds may be used in the pursuit of individual goals. Resources are not pooled. Units of exchange are determined at the lowest possible level.	Organizations exchange resources to increase each organization's abilities to achieve individual goals. Mandates or grant arrangements may provide resources. Resource needs may be satisfied by a preexisting program within an individual organization.	Pooled resources; allocation is based on balancing evolving needs of the collective group with individual constraints. Individual organizations have resources, skills, or knowledge needed to achieve collective goal. Organizational resources are allocated to support the activities of the collective unit.
Systems thinking	Organizational systems remain unchanged.	Compatible information systems can enhance coordination.	Databases are integrated to create linkages and share information between multiple layers of partner organizations.

Table 3.5 Final Variable Operationalizations: Organizational Management Construct

Variable	Cooperation	Coordination	Collaboration
Incentives	Opportunities for synergistic benefits are realized based on the desire to avoid negative impacts resulting from changes in external factors.	Grant contracts may provide funding or resource incentives to support the collective effort. Leaders identify benefits in working together and emphasize the importance of these benefits to subordinates.	Incentives are provided by the collective group and individual organizations to encourage individuals to stay involved in the collective effort.
Commitment	Work is completed as part of the regular job responsibilities conducted within the individual organization. Interests of the individual organization remain paramount. Colleagues may encourage each other to work with personnel in other organizations.	A supervisory administrative body actively encourages organizations to work together. Linkages between organizations are recognized when benefits are perceived to outweigh the costs.	Members are committed to intra- and interorganizational partners; collective interests must constantly be balanced with self-interests. Participation is justified by perceptions that the collective interest serves each organization's interests. Mutual commitment expands as organizations reciprocate collective action.
Willingness to change	Participating organizations maintain complete independence in establishing the rules, roles and responsibilities, and policies that govern the organization. An organization's standard operating procedures remain unchanged by collective efforts.	Leaders explore modifications to standard operating procedures when supporting operational goals aligned with individual organizational missions.	Partner organizations mutually adjust to the rules, roles and responsibilities, and policies collectively established to govern the collaborative unit. Changes to an organization's standard operating procedures are considered when needed to align with those of the collective unit.
Trust	Trust relationships are not required but can develop when organizations consistently share honest information.	Leaders work closely to create relationships based on trust.	Trust between organizations is necessary. Partners reinforce trust in each other by sharing information through open communication. A history of supportive interactions sustains and legitimizes relationships.

Conclusion

In sum, we find strong support for implementation of the MIM to help us sort through the different interorganizational interactions available. While we are testing in only one specific setting, the range of interactions present in the setting allows us to parse the elements in the model in such a way as to be able to evaluate the robustness of the model. Organizational interactions are complicated processes, which in turn makes theorizing about those interactions complex.

Our findings lead us to conclude that much of the extant literature is missing important differences between these interaction types, particularly those that tend to cluster around "coordination." While these interactions may exhibit some elements of coordination, the presence of other elements from surrounding interactions lends further support for a conception of these interactions as a continuum, rather than a set of discrete types. Although specific (and relative) placement on a continuum is an open question, these interactions are perhaps more usefully conceived as points on an ordinal scale, rather than simply a categorical list.

In the following chapter, we turn out attention to the ways in which these different interactions are perceived by administrators in the Coastal Zone Management Program (CZMP). By comparing these perceptions, we can also begin to unravel not only the specific elements of our proposed framework, but also the degree to which administrators engaged in program implementation perceive these relationships.

References

Boston, J. (1992). The problems of policy coordination: The New Zealand experience. *Governance: An International Journal of Policy and Administration*, 5(1). 88–103.

Gray, B. (1985). Conditions facilitating interorganizational collaboration. *Human Relations*, 38(10). 911–936.

Imperial, M. (2001). Collaboration as an implementation strategy: An assessment of six watershed management programs. *Dissertations & Thesis Full Text*, 62(02), 767. (UMI No. 3005481)

Imperial, M. (2005). Using collaboration as a governance strategy: Lessons from six watershed management programs. *Administration & Society*, 37(3). 281–320.

Jennings, E. (1994). Building bridges in the intergovernmental arena: Coordinating employment and training programs in the American states. *Public Administration Review*, 54(1). 52–60.

Jennings, E., & Krane, D. (1994). Coordination and welfare reform: The quest for the philosopher's stone. *Public Administration Review*, 54(4). 341–348.

Keast, R., Brown, K., & Mandell, M. (2007). Getting the right mix: Unpacking integration meanings and strategies. *International Public Management Journal*, 10(1). 9–33.

Keast, R., Mandell, M., Brown, K., & Woolcock, G. (2004). Network structures: Working differently and changing expectations. *Public Administration Review*, 64(3). 363–371.

Kettl, D. (2003). Contingent coordination: Practical and theoretical puzzles for homeland security. *American Review of Public Administration*, 33(3). 253–277.

Mandell, M., & Steelman, T. (2003). Understanding what can be accomplished through interorganizational innovations: The importance of typologies, context, and management strategies. *Public Management Review*, 5(2). 197–224.

Mattessich, P., Murray-Close, M., & Monsey, B. (2001). *Collaboration: What makes it work?* Saint Paul, MN: Amherst H. Wilder Foundation.

O'Leary, R., & Bingham, L. (2007). Conclusion: Conflict and collaboration in networks. *International Public Management Journal*, 10(1). 103–109.

Ospina, S., & Yaroni, A. (2003). Understanding cooperative behavior in labor management cooperation: A theory-building exercise. *Public Administration Review*, 63(4). 455–471.

Reilly, T. (2001). Collaboration in action: An uncertain process. *Administration in Social Work*, 25(2). 53–74.

Thomson, A., & Perry, J. (2006). Collaboration processes: Inside the black box. *Public Administration Review*, 55. 20–32.

Van de Ven, A., & Walker, G. (1984). The dynamics of interorganizational coordination. *Administration Science Quarterly*, 29(4). 598–621.

4 Perceptions and Formality of Administrators Implementing Coastal Resilience Policies

Analysis of Interactions during Policy Implementation

This chapter focuses on the three types of interactions within the MIM as the following research question is explored: How do administrators perceive the use of cooperation, coordination, or collaboration when working in a multiorganizational arrangement to implement policy? We employ a content analysis strategy to analyze data collected through interviews and documents. A predetermined coding scheme, aligned with the constructs of the MIM, guides data reduction. We begin our exploration of perceived use of interactions with a comparison of data across different sources.

Distribution of Interactions in Different Data Sources

The number and percentage of elements recognized within each type of interaction are identified in Table 4.1. Elements are separated based on recognition during interviews vice recognition during a review of documents. We find interactions are perceived to be highly collaborative in both interviews and documents.

Table 4.1 Summary of Elements Recognized in Interviews and Documents

Type of Interaction	# of Elements Recognized in Interviews	% of Total	# of Elements Recognized in Documents	% of Total	Combined # of Elements Recognized	% of Total
Cooperation	268	21	5	3	273	19
Coordination	293	23	43	29	336	24
Collaboration	704	56	100	68	804	57
Total	1,265	100	148	100	1,413	100

The prevalence of this type of interaction in documents is especially interesting when considering that one half of the documents reviewed are policy mandates, grant contract requirements, or a memorandum of understanding (MOU). Despite the presence of a mandate that requires organizations to work together on coastal zone issues and the presence of documents intending to formalize relationships between organizations, 65% of the elements emphasized in interviews and documents are associated with collaborative interactions.

When comparing the presence of cooperative and coordinative elements within interviews and documents, differences are evident. During interviews, cooperative and coordinative elements are mentioned in a balanced way; slightly more than one-fifth of the elements mentioned during interviews are associated with each of the two types of interactions. However, this pattern is not found in the documents reviewed. Instead, emphasis on cooperative and coordinative elements is unbalanced. While elements associated with coordinative interactions account for more than one quarter of the elements mentioned in documents, elements associated with cooperative interactions are hardly mentioned at all. The cooperative interactions that administrators perceive to occur at operational levels are simply not captured in this study's organizational documents.

Throughout implementation of the VSHP, cooperative interactions occur more often than documents suggest. In terms of the top-down/bottom-up debate, this suggests that scholars must look beyond organizational documents to capture fully the interactions that occur between organizations at operational levels. Contrary to what is assumed by much of the top-down implementation literature, a policy mandate alone cannot convey the operational patterns used to implement policy in this multiorganizational setting.

In order to explore how administrators perceive the use of cooperation, coordination, and collaboration among the organizations involved in implementing the VSHP, each interview and document is coded individually by interaction. A majority of the documents operate at the collaborative end of the interaction continuum. Two trends are evident when looking at the distribution of data. First, five of the eight documents reviewed are intended to formalize relationships between organizations involved in implementing the VSHP. These documents are the Coastal Zone Management Act (CZMA) of 1972, Executive Order Number 21, MOU for the Southern Tip Partnership, PDC Technical Assistance Grant Requirements, and

Virginia's Eastern Shore Seaside Management Plan Draft. Of these documents, administrators directly involved in the CPT and the VSHP developed the MOU and Virginia's Eastern Shore Seaside Management Plan Draft. It is interesting that of the five documents intending to formalize relationships between organizations, these two documents are the only ones that operate at the collaborative end of the continuum.

Variation of interaction types within organizational documents suggests that implementation of the VSHP embraces the combined strengths of the top-down and bottom-up approaches. Since the CZMA of 1972 and Executive Order Number 21 are explicitly co-ordinative, a purely top-down approach would prematurely limit interactions to the descriptions in these documents while ignoring factors in the local policy environment. Since the documents developed by administrators directly involved in the CPT and VSHP are collaborative in nature, they obviously see benefits in operating outside of command-and-control authorities. On the other hand, a purely bottom-up approach would fail to acknowledge the benefits associated with accountability mechanisms provided by formalized documents. An interviewee suggests that the VCZM Program staff developed the PDC Technical Assistance Grant Requirements to hold the PDCs to specific performance standards. The utility of both approaches within this research supports the study's assertion that multiorganizational implementation theory utilizes both top-down and bottom-up approaches.

Second, independent evaluations conducted by the National Oceanic and Atmospheric Administration (NOAA) support this study's finding that interactions among organizations involved in implementing the VSHP are collaborative. After a review of both documents, NOAA evaluators perceive collaborative interactions to occur between organizations. A majority of the elements mentioned in both evaluations are aligned with elements associated with collaborative interactions.

Despite the presence of documents intending to formalize arrangements between organizations, administrators in this study perceive interactions to operate beyond these mechanisms at an overwhelmingly collaborative level. Almost 70% of interviewees perceive interactions between organizations involved in implementing the VSHP to be collaborative in nature. The distribution of interview data is especially noteworthy given that the public administration literature typically associates government organizations with highly centralized and hierarchical structures; government organizations represent 60% of the organizations involved in this study. Of the government employees participating in this study, almost 90% perceive interactions to occur at a collaborative level. This suggests that government employees involved in implementing the VSHP transcend hierarchical structures within their individual organizations. In this study, the vertical linkages within individual organizations seem to be less important to the implementation network than the horizontal linkages occurring between organizations as the data leans heavily toward the collaborative end of the continuum in interviews and documents reviewed.

Overview of Perceived Interactions

The ways in which administrators perceive interactions between organizations involved in the VSHP is explored by identifying patterns within the textual data that speak to cooperation, coordination, or collaboration. Elements associated with each of the three interaction terms are present in the coding of interviews and documents. The number and percentage of elements identified in interviews and documents for each interaction are combined and identified in Table 4.2. In aggregate, elements associated with the collaborative end of the continuum are mentioned twice as often as elements associated with the cooperative or coordinative areas on the continuum.

In order to explore the perceived use of cooperation, coordination, and collaboration among organizations within the VSHP, our analysis focuses on each variable within the four constructs

Table 4.2 Percentage of Elements Explained by Type of Interaction: Interviews and Documents

	Cooperation	*Coordination*	*Collaboration*	
Interorganizational policy objective	33 (12%)	112 (33%)	264 (33%)	
Interorganzational infrastructure	61 (22%)	82 (24%)	129 (16%)	
Interorganizational procedures	96 (35%)	86 (26%)	240 (30%)	
Organizational management	83 (31%)	56 (17%)	171 (21%)	
Total within interaction	273	336	804	1,413
% of combined interaction	19	24	57	100

of the MIM. Interview and document data are combined. Within each construct, a determination is made as to where each variable falls along the continuum of interaction. From this information, an overall determination of interaction is made for each construct.

The Continuum and Interorganizational Policy Objective

The objective of the VSHP is to preserve and manage natural resources on Virginia's Eastern Shore. The policy objective is categorized as cooperative, coordinative, or collaborative based on the perceptions of administrators regarding the variables of time, difficulty, role of single organizations, and the impetus for collective action. Data collected from interviews and documents are used to place each variable along the continuum of interaction. For this study, each variable within the construct is placed at the collaborative end of the continuum of interaction. As a result, the interorganizational policy objective construct is collaborative in nature. The four variables that characterize the interorganizational policy objective are described in more detail in this section. The data suggests that the policy objective of the VSHP is located toward the collaborative end of the continuum in terms of time. More than one half of the elements identified within this variable are for the long-term nature of the policy objective. The environmental goals of the VSHP, such as habitat restoration and land preservation on the seaside of the Eastern Shore, require long-term commitments from the organizations involved. Although the VSHP was established six years ago, many of the partners began working together well before the program's creation. For example, four of the organizations involved in the VSHP own land on the Eastern Shore. An interviewee describes this commitment to land management in the following way: "Because we are owners and managers of land, our commitment is in perpetuity." It seems likely that these organizations will continue to work together, beyond the scope of the VSHP, for as long as they own land. In many instances, these organizations work together to plan the purchase and management of these properties.

In addition to the objectives of the program being described as long term, interviewees suggest that building the relationships necessary to achieve these objectives take a great deal of time. This is especially important on the Eastern Shore because agencies owning adjacent properties must collectively discuss the management of these properties. An interviewee makes this point by expressing, "It took 10 years of talking and consensus to come up with a way to view the whole ecosystem together." As the literature suggests, the building of these relationships generates interdependencies between organizations (see, for example, Bryson, Crosby, & Stone, 2006; Keast, Brown, & Mandell, 2007). The creation of interdependencies further supports long-term relationships. This assertion is supported by an interviewee who explains: "My perception is that people don't just show up to work on one project and then leave. People are pretty vested in it."

Another one half of the elements identified within the time variable are associated with the evolutionary nature of the program's objectives. As the literature suggests, relationships between organizations involved in implementing the VSHP evolve (see, for example, Mattessich, Murray-Close, & Monsey, 2001; Thomson & Perry, 2006). Organizations assume different roles depending on the project that needs to be completed and the expertise available among the participating organizations. Much like the literature suggests (see, for example, Mandell & Steelman, 2003), interactions within the implementation network are reshaped based on group dynamics and the specific task at hand. An interviewee explains the importance of this evolution within the group in the following way,

> The timeline and evolution of the process is really what is important. You couldn't just go out and take the final working relationships we all have and say that this is how this group functions and plug that in somewhere else. It really [i]s an evolution.

Throughout the discussion, the participant explains that someone facilitates the group's evolution by initiating the program, involving stakeholders, and helping the group build trust. This description aligns with the roles of a convening organization (see, for example, McNamara, Leavitt, & Morris, 2008; Wood & Gray, 1991).

In addition to the presence of a convener, the group evolves because participants individually and collectively grow throughout the duration of the project. An interview participant explains this evolution:

> I think anytime you first start a project like this it will be a little rough because there is distrust...But with time everyone learns to adjust to each other and understand each other and learn to live with each other. So I think the partnership evolve[s] positively over time.

Several interviewees suggest that the CPT initially focuses the group on easier projects, and they work their way into the projects that are more difficult. Due to the evolutionary nature of the program, participants have time to grow together. A sense of sustainability is expressed in an interview: "When September rolls around, the funding for this focal area will end, but the need for partnership does not. The need to manage the resources in a collaborative way will not end." As a result, participants are always looking for ways in which they can work with their partners on projects that grow from those directly supported by the VSHP. Another interviewee mentions that partners work within a flexible framework of dynamic processes because projects do not always work out as intended; this flexibility allows the collective arrangement to make changes as needed.

Based on participants' descriptions of the VSHP objectives, the data suggest that the policy objective is located toward the collaborative end of the continuum in terms of difficulty. More than 60% of the elements identified within this variable are for the complexity of tasks associated with the program's objectives. According to an interview discussion, the organizations within the VSHP use an "ecosystem mentality" when focusing on land management and habitat restoration on the Eastern Shore. As a result, many people and organizational entities are involved in this regional approach. Of the participants interviewed, more than 67% indicate that the objectives of the program are highly complex. Many attribute this complexity to the nature and scale in which the program is trying to resolve environmental issues. A high level of difficulty can be seen when looking at a technique the Division of Natural Heritage uses to map Phragmites on the Eastern Shore. Phragmites is an invasive plant species that disrupts the natural landscape. This technique is explained during an interview:

> We've mapped the entire seaside of the Eastern Shore. For example, we've developed a technique using low elevation flights with helicopters and global positioning systems to map Phragmites at a really high scale—a high resolution with a lot of precision and accuracy with this mapping. We are mapping tiny patches of Phragmites and essentially doing a census of all the Phragmites on the seaside of the Eastern Shore. This technique didn't exist before.

Projects encompass highly varied tasks that involve different stakeholders, raise different issues, and focus on different goals. Some of the projects include shellfish restoration, shore bird habitat protection, Phragmites control, and the development of ecotourism. As suggested by the literature (see, for example, Keast, Brown, & Mandell, 2007), collaborative interactions are used in the VSHP to address highly complex problems.

Another 30% of the elements identified within the difficulty variable are associated with the program's objectives addressing a situation of crisis. With great focus placed on the Chesapeake Bay, the Eastern Shore historically receives little attention. A common theme among interviewees is that the seaside of the Eastern Shore is forgotten. This point is explicitly expressed by an interviewee in the following manner: "The Seaside is always forgotten because of the Chesapeake Bay. It is one area that is so rich in natural resources but it can fall through the cracks." In and of itself, a lack of attention may not be grounds for crisis. However, the seaside of the Eastern Shore is a critically important environmental area. An interviewee describes the area as "one of the world's most important biospheres." Another participant agrees: "The Eastern Shore is a jewel. You have a suite of wildlife resources found on the Eastern Shore that are not only valuable from a scientific and conservation standpoint but are [also valuable as] major economic and recreational resources." Grounds for crisis arise because there is an environmentally significant area faces severe development pressures compounded by economic stress and a lack of attention.

Based on the textual data collected through interviews and documents, the policy objective is located toward the collaborative end of the continuum in terms of the roles single organizations play. About one half of the elements identified within this variable are associated with interdependencies between organizations. It is important to recognize the way in which interdependence applies to interactions in this study. The literature suggests that collaborative interactions occur when single organizations cannot resolve a problem individually (see, for example, Wood & Gray, 1991) and each organization becomes one piece of a larger system (see, for example, Mandell, 1994). While interview responses indicate that participants perceive their organizations to be interdependent with those of a larger system, this study does not necessarily support the literature's nuances.

In this study, organizations are not perceived to be unable to function independently. To the contrary, interviewees suggest that their organizations can operate individually to make some differences on the Eastern Shore. An interviewee explains that each organization independently makes technical recommendations, as-sumes management roles, and exerts legal authorities. Interdepen-dence comes into play when interviewees discuss the magnitude, scope, and successes of what they accomplish when working to-gether. An interviewee clarifies this distinction of interdependence in collaborative interactions:

> Even though each agency can do its own piece fairly well on its own, the objectives of the VSHP are certainly much broader than any one of the agencies. Even if we are able to work mostly independently on our little piece of it, it is just a piece.

Resource administrators and operational personnel alike indicate that they look to find ways to tie their organizations and research together because they are able to accomplish more by doing so. This finding is supported in research conducted by Keast, Brown, and Mandell (2007). Based on data collected in interviews and focus groups with policy makers and practitioners in the service arena, their research indicates that collaborative interactions are employed when organizations search for ways to "achieve greater efficiencies of scale and outcome" (p. 18).

The data suggest that the policy objective of the VSHP is located toward the collaborative end of the continuum in terms of impetus for collective action. More than one half of the elements identified within this variable indicate that a convener brings organizations together, they have complementary interests, or they can better achieve the desired goal when working together. First, organiza-tions come together to implement the VSHP because the VCZM Program plays an important role in convening the group. As is sug-gested by the literature (see, for example, Bryson, Crosby, & Stone, 2006; McNamara, Leavitt, & Morris, 2008), the VCZM Program serves as a mechanism to encourage interactions between organi-zations and establish the collective arrangement. The importance of identifying a mechanism to bring organizations together is ac-knowledged during an interview: "The key is having something to bring these organizations together. Otherwise they will work to-gether where it benefits them." Another interviewee suggests that the convening role of the VCZM Program is especially important because each organization is individually busy and involved in many other projects. Due to their expertise in facilitating relation-ships, the VCZM Program staff has high levels of credibility with their partners. This credibility is frequently mentioned in interviews and documents. The literature indicates that credibility is an essential component in influencing the organizations to work together in an arena where formal authority is nonexistent (see, for example, Wood & Gray, 1991).

Second, organizations come together because they identify complementary interests. An interviewee expresses this point during an interview: "Our objectives are clearly a subset of the overall VSHP. The VSHP covers a lot of bases well beyond the narrow mission of our division. But it is quite aligned." The VCZM Program also plays an important role in helping organizations recognize when they have complementary interests that can be better served by working together rather than alone. An interviewee explains the involvement of the VCZM Program as follows:

> The interests of those agencies have significant overlap in those areas that are particularly ecologically important. In their daily work they often don't think about that overlap, they only think about what they are doing. My perception is that when they get together under the auspices of the Coastal Zone Management Program they tend to look more at how they can work together.

Interviewees indicate that they perceive their own organizational goals to be furthered by establishing partnerships with other organizations on the Eastern Shore. This desire to serve individual organizational interests while also meeting collective interests is supported by the literature (see, for example, Thomson & Perry, 2006).

Third, organizations work together to implement the VSHP because personnel perceive that they can better achieve the desired goal. Funding concerns appear to drive this perception as interviewees frequently mention a lack of resources and tight budgets. Like many public organizations, those involved with the VSHP have fewer resources to face increasingly complex problems. An interviewee suggests that scarce resources bring organizations together. "We have a huge mandate and little resources to accomplish it with. So we have a vested interest to work together." Leveraging resources and money help organizations achieve their goals. The need to leverage resources is described by an interviewee as follows:

> The job that needs to be done is bigger than any one agency. And things like the Virginia Seaside Heritage Program give you a vehicle for everyone to work together...to get in the same

car and to get to the same place with somebody else providing the fuel—[the Virginia Coastal Zone Management Program].

The need to leverage funds is also mentioned in tandem with the matching requirement on funding distributed by the VCZM Program. An interviewee suggests that multiorganizational interactions are a necessity for implementing the VSHP:

> Even if we ha[ve] all the expertise in the world we could never afford to take on the entire program because we couldn't come up with the match money to do it. VIMS [Virginia Institute of Marine Science] could come up with the match money to do it but they probably wouldn't have all the resources in place to say that they could do it. And by resources I mean personnel, background, and physical plant resources such as boats, labs, and computer space to do the entire job. So it is a program that needs to be done as a partnership.

The Continuum and Interorganizational Infrastructure

The VSHP is comprised of a network of organizations that work together to preserve and manage natural resources on Virginia's Eastern Shore. The infrastructure is categorized as cooperative, coordinative, or collaborative based on the perceptions of administrators regarding the variables of design, formality of the agreement, organizational autonomy, policy authority, and key personnel. Data collected from interviews and documents are used to place each variable along the continuum of interaction. Variables within this construct are placed in different areas along the continuum. Despite this variation, the interorganizational infrastructure construct is collaborative based on the elements emphasized throughout interviews and documents. The five variables that characterize this construct are described in more detail in this section.

Based on participants' descriptions on the ways in which the multiorganizational arrangement is generated and structured, the data suggests that the infrastructure is located toward the collaborative end of the continuum in terms of design. Over 40% of the elements identified within this variable are associated with the new program structure used to implement the VSHP. The creation of a new program structure is consistent with the literature (see, for example, Mattessich, Murray-Close, & Monsey, 2001) and suggests that participants identify with the horizontal linkages used to implement the VSHP.

Two types of horizontal structures are used to establish linkages within the multiorganizational arrangement. The CPT is one type of horizontal structure used to implement the VSHP. Many interviewees describe this group as providing the leadership and structure needed to bring the network of participants together. An interviewee explains the role of the CPT:

> Each time you look at a project, it is a collection of partners that have all come together. And I don't know if those partners would have necessarily worked as well together if there hadn't been a structure to bring them together.

This governing body is comprised of representatives from state and local governments. Participants of this group make programmatic decisions to guide the overall direction of the VCZM Program. In addition, they make decisions regarding the distribution of grant funds.

The executive steering committee is a second type of horizontal structure used to implement the VSHP. This group is comprised of operational personnel who have field-level expertise and are responsible for managing projects on the Eastern Shore. Their expertise is widely acknowledged by interviewees holding positions on the CPT. During interviews, many CPT members recognize that personnel at the operational levels have 20–30 years of experience in studying these ecosystems, and this knowledge helps them make sound decisions in terms of project operations.

These findings indicate that multiorganizational implementation requires the development of horizontal connections between organizations in addition to vertical connections within organizations. The use of the CPT and executive steering committee facilitates the involvement of two levels of personnel from each state agency—resource administrators and operational project leaders. Representatives on the CPT typically supervise the project leaders on the executive steering committee. In addition, multiorganizational arrangements may benefit from horizontal connections at more than one organizational level. Operations within the VSHP appear to run smoothly because resource administrators are horizontally linked with one another while operational personnel from the same agencies are also horizontally linked with one another. It is through these linkages that resource administrators and operational personnel become aware of the expertise within their own organizations and in other organizations.

Textual data suggests that the interorganizational infrastructure construct is located in the coordinative area of the continuum based on the formality of the agreement between organizations involved in the VSHP. More than one-third of the elements identified within this variable are attributed toward contractual agreements formalizing relationships. This is not surprising considering that the VCZM Program awards one half of a million dollars in grant funds annually to partners involved in the VSHP. As a result, roles and responsibilities are largely determined by stipulations within the grant. An interviewee describes the formality of agreements between organizations as "partly grant driven." Another interviewee says, "[Roles] are usually based on our specific mandates for regulatory initiatives or controls." Grant contracts are used to identify projects suitable for collective action and formalize relationships between participants implementing the VSHP. Contracts are used in a similar manner in the research conducted by Jennings and Krane (1994). In collaborative interactions, hierarchical forms of accountability are nonexistent because participants are considered to have equal status. Therefore, situations involving the distribution of funds require the presence of formal accountability mechanisms. Since these mechanisms are not inherently present in collaborative arrangements, grant contracts serve this purpose.

Based on analysis of interview transcriptions and documents, the infrastructure of the VSHP is located toward the collaborative end of the continuum in terms of organizational autonomy. More than one half of the elements identified within this variable are attributed to organizations not being autonomous. A common theme in the data is that organizations can function independently in their day-to-day operations, but overall operations are inherently intertwined with those of other organizations implementing the VSHP. An interview participant suggests that organizations benefit by working together: "Everyone could do their job on their own. They are able to do a much better job by working together." Interviewees recognize that an ecosystem approach to managing resources on the Eastern Shore requires them to work with other organizations in order to better meet the program's diverse objectives. Discussion during an interview reveals why organizations are not autonomous when implementing the VSHP:

> The goals of the program are pretty broad so no one agency can do it themselves. You have to have that mix of expertise and disciplines to cover the bases of all the different resources that are on the Eastern Shore...[T]here is so much to do that you need a lot of different hands and you need a lot of different

expertise because of the fact that no one organization has suffi-
cient capabilities and expertise in all the different disciplines to
address the broad range of resource issues that present themselves on a place like the Seaside.

Recognition that collective efforts are needed to accomplish the
totality of the task aligns with the research (see, for example, Mandell, 1994). Interdependencies are captured during an interview:

> Certainly at the state level we rely on each other quite a bit
> because everyone has their own piece of it and if somebody's
> piece doesn't get done than that has a major impact on the total
> objective of the VSHP. It requires all organizations that are in-
> volved to finish the elements they agree to. So the success of the
> total project is dependent on each of the elements.

Multiple interviewees convey that each organization addresses one
piece of the Eastern Shore's larger ecosystem, and each piece impacts the larger system.

Interviewees emphasize that their organizations are individual
pieces of a larger ecosystem; this emphasis suggests that collaboration between organizations involved in implementing the VSHP
occurs in a specialized way. Other research on collaboration in
the environmental arena supports this theme (see, for example,
McNamara, Leavitt, & Morris, 2008). Organizations utilize specialized expertise while working on various projects pertaining to
natural resources on the Eastern Shore. It likely takes great understanding of each organization to align these independent specializations in ways that meet the goals of the VSHP. The role of the
VCZM Program staff in aligning organizational specializations is
described by an interviewee:

> It's like being a conductor of a symphony. You have your differ-
> ent instruments and you know what their specialties are. So you
> figure out the right time to bring them in and hopefully it comes
> together in one nice piece of music.

In this sense, collaborative interactions are purposive to the extent
that the convener brings together a group of organizations with the
specializations needed to carry out the program's objectives.

The data suggests that infrastructure is located toward the cooperative and collaborative ends of the continuum in terms of

policy authority. There is only a one-point difference between the elements identified as cooperative and those identified as collaborative. In addition, almost one half of the elements identified within this variable are associated with each of these two interactions. This is important to recognize because in this study the variable operates at both ends of the continuum. Although the data may appear contradictory, these seemingly dichotomous views may be attributed to participants focusing on different levels of policy decision making when answering the interview question, the sectors involved in the program, and the design of the CPT.

In the VSHP, local governments and federal/state agencies largely represent the public sector. These organizations have responsibilities mandated through specific legal authorities. Personnel representing these organizations may recognize a resource that would benefit from the development of enforceable policies and gather data to support a particular policy change, but policy decisions ultimately occur through a political process. This emphasis on organizations independently following preexisting policies is aligned with the cooperation literature, which suggests that organizations retain separate identifies and control resources individually when working together (see, for example, Keast, Brown, & Mandell, 2007). An interviewee expresses a need for each agency to follow its legal authorities: "For the Eastern Shore group, each agency has to be very clear in voicing their concerns and their legal authorities or restrictions...There may be certain things that an organization just cannot do."

On the other hand, these same participants also indicate that the CPT can make operational policy changes within the scope of each agency's legal authorities. Two types of decisions are made by the personnel on this team. First, programmatic decisions occur at the operational level. An interviewee mentions: "Agencies do develop collective policies to guide operations." It is through the CPT that decisions are made regarding commitments to projects and allocation of resources. Another interviewee indicates that the partners work together to identify the program's focal areas. "We all work together as a team and we make decisions as a team as to where the focal area will be."

Second, decisions are made regarding the focus of future research. These decisions are typically guided by a desire to provide state policymakers and citizens with the information needed to make sound policy decisions regarding land-use on the Eastern Shore. In situations where participants of the VSHP desire policy changes that occur outside their programmatic boundaries, interviewees emphasize that they can influence the political process through research and the communication of findings. The potential to influence this process is explained during an interview:

> The policies that we develop are probably one of two things. First, deciding as a group what area we should go in for enforceable policies...Most of the other policy development has been in terms of seeing a policy need and developing the information behind it and a recommendation on what policy should be and presenting that to the appropriate agencies for them to take it to the appropriate channels.

Although these efforts do not directly change policy, it is through this information process that interviewees believe they indirectly impact enforceable policies.

This distinction between programmatic policy authorities and more general policy authorities are not necessarily clear in the literature. The literature suggests that organizations jointly develop rules and procedures to guide the collective unit (see, for example, Bryson, Crosby, & Stone, 2006). Although this assertion is supported by the findings in this study, it is important to consider that these developments occur within the context of the programmatic level. This may provide a more realistic application of the policy authority variable during collaborative interactions between public organizations who are guided by specific legal authorities.

Based on the data collected, the infrastructure of the VSHP is located toward the collaborative end of the interaction continuum in terms of key personnel. Almost 20% of the elements within this variable speak to the involvement of the convening organization in proposing policies and rules for the collective group to consider. An interviewee describes the role of the convener by saying, "Money brought everyone to the table and good leadership brought everyone together." The VCZM Program represents the staff element involved in all operations of the collaborative group; the presence of a staff element is also identified in Agranoff's (2006) research on public management networks.

The findings from this study align nicely with the literature's discussion on champions and sponsors (see, for example, Agranoff, 2006; Mandell & Steelman, 2003). In terms of implementing the VSHP, NOAA acts as a sponsor by providing the authority and resources to legitimize the implementation network. The VCZM

Program staff act as the champions for the implementation network because they sustain interactions with their needed expertise. This theme emerges throughout numerous interviews. An interviewee reveals the presence of a champion within the multiorganizational arrangement: "[T]he Coastal Zone Management Program has been a champion and really got the project going and got people involved." Another interviewee suggests: "From day one, it was always put forth that the reason this is possible is because it is a regional approach. We have to all be working together." The results from this study align with Agranoff's (2006) research on public management networks. In both cases, champions play a significant role in encouraging organizations to support the collective arrangement.

In addition, more than 15% of the elements within the key personnel variable speak to the adaptability of membership, roles, and responsibilities. An interview participant explains why organizations change roles: "Certain groups are involved in specific projects depending on their expertise." For example, the Virginia Institute of Marine Science and the Virginia Marine Resources Commission are primarily involved in oyster restoration. The Nature Conservancy and the Center for Conservation Biology are primarily involved in avian research. On the other hand, the Department of Conservation and Recreation is focused on Phragmites control and providing opportunities for ecotourism.

Although these examples only represent a fraction of the organizations involved in the VSHP, a broader look reveals a similar pattern—each organization has a niche within the bigger group. An interviewee suggests that flexibility is needed to work on the diverse tasks associated with the VSHP. "You need some flexibility because nothing ever works completely as planned. You want to be able to retool and regroup in order to have a dynamic process." Despite a need for the group to adapt to the task at hand, dynamics within this implementation network stabilize through the specialized nature in which the organizations come together. Although different tasks require the expertise of different organizations, it seems that particular subgroups repeatedly work together.

The Continuum and Interorganizational Procedures

Participants within the VSHP indicate that processes within the arrangement support and sustain relationships. Procedures are categorized as cooperative, coordinative, or collaborative based on the perceptions of administrators regarding the variables of information sharing, decision making, resolution of turf issues, resource allocation, and systems thinking. Data collected from interviews and documents are used to place each variable along the continuum of interaction. Each variable within the construct is placed at the collaborative end of the continuum. As a result, the interorganizational procedures construct is collaborative in nature. The five variables that characterize interorganizational procedures are described in more detail in this section.

The data suggests that interorganizational procedures are located toward the collaborative end of the continuum in terms of information sharing. More than 40% of the elements identified within this variable are attributed to open and frequent communication and the willingness to share information. Regular meetings among individuals involved in implementing the VSHP and routine communication among personnel working at the operational level facilitate collaborative interactions between organizations. The benefits of communication are described by an interviewee in the following manner:

> I've been in Virginia doing this now for just over 10 years and this certainly enable[s] me to learn who many of the other players are in the coastal area working in natural resources. I've learned a lot about who they are, how they operate, who I can count on, and who not to count on.

It is through regular meetings of the CPT that partners openly communicate with one another as they discuss the direction of the VSHP, identify what their organization can provide to the collective group, and learn more about the other organizations involved in the program.

In addition, it is common for personnel representing the partnering organizations to communicate in the course of their daily operations. An interviewee describes daily communications among organizations implementing the VSHP: "There is so much routine contact here that when it comes time for all the partners to come together the only hard part is figuring out a date." These communication linkages are further strengthened by long-standing relationships and geographic proximity among partners on the Eastern Shore. Another interviewee suggests, "It is completely common to pull up to a boat ramp and see several partners. And you stop and talk." As is consistent with the literature (see, for example, Thomson & Perry, 2006), open and frequent communications between partners involved in implementing the VSHP reduce information asymmetries.

Furthermore, a willingness to share information between organizations facilitates collaborative interactions between partners implementing the VSHP. As organizations enhance their understanding of one another, they become increasingly willing to share information. During interviews, a participant suggests that this willingness to share information with one another "helps create a real scientific community rather than a group of scientists." In sharing information on restoring coastal habitats, replenishing aquatic resources, and promoting sustainable economic activities on the seaside of the Eastern Shore, the collective group is better able to employ an ecosystem approach. A sense of understanding within the network is described during an interview: "Now everybody understands each other's work so completely that they realize that none of their projects is more important than any others. And they are looking for opportunities to find ways to help others." As organizations focus on projects that address one piece of the larger ecosystem, a willingness to share information allows them to become more knowledgeable in areas that address interrelated pieces of the ecosystem. Developing a knowledge base through information sharing and understanding is supported by literature (see, for example, Imperial, 2001; Keast, Brown, & Mandell, 2007).

Stability among partners involved in the VSHP contributes to this willingness to share information. An interviewee describes the development of stability within the multiorganizational arrangement: "It is the same group of organizations that come together on a regular basis to identify what the resources needs are, to see what people and money each organization has, and to pool the resources to conduct conservation." In many instances, the VCZM Program staff encourage organizations to share information. This point is further explained during an interview: "The VSHP provides platforms for people to talk in ways and at levels of intimacy that averts problems more than it creates them."

Interorganizational procedures are located at the collaborative end of the continuum in terms of decision making. Elements within collaborative decision making are identified 12 times more often than the elements within cooperation or coordination. More than 80% of the elements identified within this variable are attributed to participative decision making based on consensus and compromise. This process is described by an interview participant: "Decision making is a collegial process. There are a lot of prioritizations to be made. It is an open, roundtable discussion. And we try to come to consensus on what the priorities will be." Another interviewee agrees: "The discussion happens with all the partners sitting at the table." Much like the literature suggests (see, for example, Agranoff, 2006; Mandell & Steelman, 2003), consensus and compromise are an important part of the process. It is common for interview participants to describe the process as an open discussion. Based on the following statement from an interviewee, participants perceive themselves to have great input into decisions:

> The whole group discusses and decides the priorities. Once the group as a whole sets the priorities, the partners most applicable to that project talk amongst themselves about how to carry it out. Small groups form around particular projects. All the [needed] partners [are] at the table.

It is clear during interviews that participants consider themselves equal stakeholders when it comes to making decisions.

Interorganizational procedures are located at the collaborative end of the continuum in terms of resolving turf issues. More than one half of the elements within this variable are associated with maximizing common ground and recognizing incongruent demands between individual organizations and the collective group. Therefore, turf issues are resolved in two ways. First, participants within the VSHP focus on maximizing common ground in order to minimize turf issues. Although each organization may have a different interest in protecting the seaside of Virginia's Eastern Shore, interview responses suggest that organizations maximize common ground by focusing on the needs of the resource. An interviewee expresses recognition of this common ground: "It's all different roads leading to the same destination." A common goal unites the organizations implementing the VSHP. This unity is explained in an interview: "When the bottom line is the protection of the resource, and this is what you are focused on, I think it is easier to resolve these issues." As personnel from different organizations agree on the need to protect the resource, interviews also realize no organization can accomplish the goal individually. An interviewee acknowledges the necessity for organizations to work together: "In a lot of cases, we've realized that we need each other. And nobody has the resources we used to have so we can't afford to fight with each other."

Second, participants within the VSHP minimize turf issues by recognizing that there is potential for incongruent demands between individual organizations and the collective group. The potential for incongruent demands is also recognized in the research conducted by Thomson and Perry (2006). The need to balance competing demands is acknowledged by an interview participant:

> There is a balancing act between the interests of the individual organization and those of the collective group. And the collective has to recognize the mandates and the limitations of the individual partners. You strive to identify the things that everyone can support and then you continue the hard work with some of the tougher issues. These can be addressed; it just takes longer.

In balancing the interests of individual organizations and the collective arrangement, turf issues do not impact the network of organizations involved in the VSHP. This finding aligns with Agranoff's (2006) research on collaborative networks involving federal, state, and local governments. The literature indicates that conflicts between organizations may be resolved by adjusting policies and procedures (see, for example, Mattessich, Murray-Close, & Monsey, 2001). In this study, resolution of turf issues does not appear to come from policy or procedural changes. Instead, findings suggest that there are two other explanations for the lack of turf issues between the organizations involved in implementing the VSHP.

First, interview responses indicate that organizations within the implementation network focus on understanding the different perspectives and concerns of other participating organizations. "It is not enough to accurately hear what other [people are] saying, you actually have to understand why they are saying it, what their perspective[s] [are], and what they really need." Interviewees indicate that they spend great amounts of time discussing what programs to pursue and how to implement them. When problems arise, they also spend a great deal of time resolving them. Significant emphasis is placed on identifying common opportunities that involve projects deemed valuable by a majority of organizations.

Second, interview responses suggest that a lack of turf issues may be somewhat predetermined based on the organizations identified to implement the VSHP. The organizations involved in the VSHP generally have specialized roles based on distinct mission areas. These organizations are brought to the table because their mission areas are tangentially related and focus on the seaside of the Eastern Shore. However, mission specializations help minimize conflicts among partners because the allocation of grant money and determination of project involvement are often based on the need for a particular expertise. Therefore, conflict is minimal because organizations do not need to compete for the same funds or project involvement. This finding suggests that turf issues can be avoided to some extent based on the organizations brought to the table. Conveners should give considerable thought to identifying the specializations needed to implement a particular objective prior to bringing the organizations together. As a result, the outcome of collaborative interactions may be related to how well the convener accomplishes this task.

Interorganizational procedures are located at the collaborative end of the continuum in terms of resource allocation. The number of times that elements within this variable are mentioned increase along the continuum of interaction in a linear fashion from cooperation to collaboration. Almost 30% of the elements within this variable are associated with resources provided through grant contracts. This element aligns with coordinative resource allocation and is mentioned more often than any other element. An interviewee explains the criticality of a stable funding stream in sustaining partnerships within the VSHP:

> When you have money, you can do some things that you could never do. And you can get people to work with you in ways that they would have never worked with you before. If you can start paying for things then people start chipping in their time. The matching aspect a lot of organizations are capable of but if there isn't a funding source to drive the whole thing then [interaction] is harder to come by. That has been the story of this program.

A majority of interviewees suggest that the success of the VSHP is possible because of the funding stream provided by the VCZM Program. An interviewee discusses the impact of funding: "The funding and the possibilities that the funding creates for action is what makes the progress possible." Interviewees, employed by public organizations, indicate that these additional funds directly contribute to the scope of the work they are able to accomplish.

The structure of the grant process generates opportunities for organizations to pool resources around the funding stream. Through a pooling of resources, organizations implementing the VSHP move toward the collaborative end of the continuum in terms of resource allocation. An interview participant explains that much of the grant money distributed by the VCZM Program requires organizations to have a one-to-one match with nonfederal monies, "The match requirement lends itself nicely to pooling resources. Often, the money is matched with time and personnel." It is through the matching requirement that organizations identify opportunities to leverage resources. An interviewee mentions: "We go out to other organizations and line up funds that will benefit the Seaside." Leveraging resources enhances the power of the implementation network. An interview participant makes this point during discussion by expressing, "Leveraging people's resources is really the best power of the VSHP. They are asking for a one-to-one dollar match so you have to have funding from elsewhere."

When elements within the resource allocation variable are aggregated by interaction, it is clear that resource allocation occurs at a collaborative level. Although this pooling of resources occurs, it may occur in a different way than the literature suggests. An interviewee explains the applicability of pooled resources to the VSHP:

> They are pooled to the extent that everybody contributes. They are not pooled to the extent that you donate a certain amount of time and somebody else decides how that time is spent...[I]f I am going to commit my time or my staff's time to a specific task, than it [has to be] within our mission and my responsibility.

In some instances, organizations leverage resources because their individual projects align with the goals of the VSHP. Several interviewees describe this process as "piggybacking." In these instances, the VSHP benefits from the completion of additional projects while individual organizations benefit from additional funds.

Other times, the pooling of resources occurs because organizations have access to funds that other VSHP participants do not have access to. As these funds are identified by organizations involved in the VSHP, they often apply them to projects that align with the initiatives of the multiorganizational arrangement. It is mentioned during an interview that the cost of land on the Eastern Shore often requires organizations to pool various funding sources in order to purchase a piece of property.

> The most recent acquisition project involved a piece of land contiguous to the Fish and Wildlife refuge at the Southern Tip. We came together as a group to figure out what pots of money might be available to buy that piece of land...The money

is coming from all different pots because no one source has enough cash to pay for it all.

In these situations, resources are generally pooled in the sense that one group may be able to push an initiative forward in a way that another organization may not be able to. For example, a nonprofit like The Nature Conservancy (TNC) does not face the same budget constraints that government organizations face, and they are able to use their money in ways that public organizations cannot. Therefore, TNC often spearheads land acquisition on the southern tip of the Eastern Shore because the organization can quickly allocate the funds and purchase the desired property. Due to the bureaucracy within public organizations, they are unable to operate at the same speed. As a result, TNC often purchases land initially and then works with various federal/state agencies to determine who will repurchase and manage it.

Furthermore, nonfinancial resources are pooled in a specialized way. Interviewees frequently mention that their organizations have resources needed to achieve the program objectives and that these resources are allocated to the collective group. For example, TNC has access to volunteers and great expertise in terms of land acquisition and bird habitat management. The Eastern Shorekeeper provides informal enforcement to ensure restored areas remain undisturbed. Academic institutions, such as The College of William and Mary and the University of Virginia, provide data that is used to advocate specific management decisions. The VCZM Program staff members are experts in grant management and environmental facilitation. These are just a few examples of the ways nonfinancial resources are pooled in a specialized way. While Agranoff's (2006) research suggests that government managers contribute resources to collective groups, this study suggests that the pooling of resources occurs in a specialized way in which each organization retains control of the resources they provide to the collective group.

Interorganizational procedures are located at the collaborative end of the continuum in terms of systems thinking. Two-thirds of the elements within this variable are associated with integrating information systems to foster linkages between organizations. Much like Imperial's (2005) research, organizations within the VSHP use interagency databases to make information widely accessible to all participants. The coastal geospatial and educational mapping system (GEMS) is funded by the VCZM Program and often cited by interviewees as a useful web-based tool. An interviewee explains this tool, "Information is housed in one site—the Coastal GEMS program. This helps keep the organizations aware of what is going on so we know what the other organizations are doing." Organizations can view land-use and resource management information through this program. In some instances, the VCZM Program requires organizations receiving grant funds to produce a data layer to add into Coastal GEMS. Several interviewees explain that this approach encourages organizations to support the database and increases organizations' willingness to share information. In addition to developing and maintaining Coastal GEMS, the VCZM Program helps organizations identify common needs and see the importance in sharing information.

The Continuum and Organizational Management

Participants within the VSHP indicate that behaviors within and between member organizations support the multiorganizational arrangement. Organizational management is categorized as cooperative, coordinative, or collaborative based on the perceptions of administrators regarding the variables of incentives, commitment, trust, risk taking, and willingness to change. Data collected from interviews and documents are used to place each variable along the continuum of interaction. Variables within this construct are placed in different areas along the continuum. Despite this variation, the organizational management construct operates at a collaborative level based on the aggregation of elements mentioned. The five variables that characterize organizational management are described in more detail in this section.

Organizational management is located in the coordinative area of the continuum in terms of incentives. More than one-third of the elements identified within this variable are attributed to the provision of funds through grant contracts. An interview participant suggests that money is the carrot that initially brings people to the table.

> The VSHP [i]s an effort to get all the researchers in a particular region working together and to fund them to a level so they c[an] achieve significant success in a relatively short amount of time and not have to spend all their time chasing money.

Contrary to Jennings' (1994) research, incentives attached to financial provisions are emphasized twice as often as incentives linked to leadership support. Incentives based on funding are discussed during an interview:

> Having sufficient funding allow[s] us to work better together because we d[o]n't have to fight each other. At the end of the day we [a]re able to look at our priorities and look at our projects, and since we [a]ren't needing to compete for the money, we [a]re able to work better together.

An emphasis on funding rather than leadership support may be explained by the independence afforded operational personnel and discretion to determine their involvement in the implementation network. In some instances, interviewees seem largely removed from the bureaucracy typically associated with government agencies. While not emphasized as an incentive for participation, findings from this study acknowledge the time and resources middle-level organizational leaders commit to the VSHP through their involvement on the CPT. Perhaps participants are given increased levels of independence and discretion because of support from leaders behind the scenes.

Organizational management is located at the collaborative end of the continuum in terms of commitment. More than one half of the elements identified within this variable are attributed to the need to balance individual and collective interests, the collective interest serving the organization's individual interests, and the extent to which relationships are reciprocated throughout the arrangement. As indicated in the literature (see, for example, Mandell & Steelman, 2003; Thomson & Perry, 2006), interviewees suggest that there is a need to balance individual and collective interests. While organizations involved in implementing the VSHP are committed to the Eastern Shore, they are also committed to their individual organizations. An interviewee suggests that interactions occur when these interests intersect: "When we see our missions cross, we work together. But, we also work independently." Another interviewee adds, "Each agency participates within the lines of their mission. So if they don't feel a direct connection with their mission it would not be worth their time to continue to participate." The literature also suggests that competing interests between the individual organization and collective group may create tension (see, for example, Mandell & Steelman, 2003; Thomson & Perry, 2006). High levels of tension are unfounded in this study, and this may be because the convener brought specific organizations to the table based on an identified need for particular expertise. When convening the group, the VCZM Program staff takes into account a need to balance organizational interests.

In addition, this study indicates that commitment within the VSHP is located at the collaborative end of the continuum because the missions of individual organizations advance through the collective group. As the literature acknowledges (see, for example, Keast, Mandell, Brown, & Woolcock, 2004; Thomson & Perry, 2006), collaboration occurs between partners implementing the VSHP because they can resolve complex environmental problems without diminishing their commitments to individual organizations. Interviewees indicate that their organizational interests are met while working together. These efforts are described by an interview participant: "We are meeting our objectives for resource management and land conservation but we are also furthering the whole effort." An interviewee acknowledges that individual organizations benefit by working together.

> Beyond the funding opportunities, it gives us the opportunity to do work that we otherwise would not be able to do and achieve a part of our mission that would otherwise not be possible. It is an opportunity to be successful in a way that would be impossible otherwise. It creates opportunities to work with other agencies in a way where the whole is greater than the sum of the parts. The benefits transcend your first expectations because of the ideas that are generated.

The VCZM Program staff gives much thought to convening a group of organizations whose individual interests are served by the collective interest. In addition, they ensure organizations understand how these interests align.

Furthermore, relationships between organizations are largely reciprocated within the implementation network of the VSHP. High levels of reciprocation are seemingly enhanced by personal commitments from the people involved in implementing the VSHP. It is evident during interviews that participants feel personally connected to the work they do. An interviewee explains this connection: "All the people within these organizations care so deeply about the place—every one of them. I can't think of one person in that VSHP partnership that I would say is not just deeply and personally committed to saving this place and making it better."

Organizational management is located at the cooperative and collaborative ends of the continuum in terms of willingness to change. There is only a one-point difference between the elements within cooperation and those within collaboration, and more than one-third of the elements identified within this variable align with both types of interactions. This is important to recognize because in this study the variable operates at both ends of the continuum. Perhaps these seemingly dichotomous views can be attributed to the extent to which the government sector is involved in the program and the design of the CPT. A presence of cooperative elements may be explained when considering that more than 70% of interview participants are government employees. Government agencies are mandated by specific legal authorities and responsibilities; interviewees recognize that their organizational policies are independently established and must remain unchanged by the collective arrangement. This independence is acknowledged: "[Organizations] have [their] own internal guidelines and state code." If changes are in opposition to legal authorities or regulations, organizations are not able to make changes. Therefore, the cooperative element within the willingness to change variable may be attributed to the strong presence of government organizations and the separate legal authorities that guide each organization.

On the other hand, collaborative elements within the willingness to change variable may be attributed to the presence of the CPT. Interviewees indicate that their organizations are willing to consider changes to the policies or procedures that govern project operations within the field. For example, interviewees mention that changes to comprehensive plans, master plans, and research agendas are made as a result of activities within the collective arrangement. According to an interview participant,

> As long as it [i]s something that we have control over as an agency, that isn't mandated from somewhere above the agency, and it is reasonable we w[ill] try to work with the Coastal Zone Management Program and the Coastal Policy Team to change policies or procedures.

Another interviewee indicates that his division is always looking for "the innovative way of doing something." Discussions during interviews indicate that the research being conducted on the seaside of the Eastern Shore can lead to more informed management and resource practices. Therefore, organizations are willing to make changes to improve operations provided that it is within the boundaries of their legal authorities and regulations.

Organizational management is located at the collaborative end of the continuum in terms of trust. More than 85% of the elements identified within this variable indicate that there is a history of supportive behavior and long-standing relationships between the organizations involved in the VSHP. "A history of supportive interactions sustains and legitimizes relationships" element is mentioned more often than any other within the entire model. An interviewee explains that many of the players currently involved in implementing the VSHP started working together almost 20 years ago to protect the Mid-Atlantic migration corridor—a piece of property on the southern tip of the Eastern Shore, which is an important stopover for migratory songbirds traveling from South and Central America to Canada. Another interviewee adds that many of the organizations have "worked together long before the VSHP came along." Twenty years later, the organizations and people representing these organizations still interact with the VCZM Program in significant ways. Discussions during interviews suggest that many personnel spend their entire careers on the Eastern Shore. "The secret of success [is] the continuity of the personnel over time." Another participant expresses agreement: "And it [is] a very stable group. The partners that were in it from the beginning are largely still in it." As a result, these organizations seem to have a deep understanding of the resource and other organizations involved.

In addition, a common theme among interviewees is that trust within the multiorganizational arrangement operates at the collaborative end of the continuum because organizations work together in a variety of ways. Discussion during an interview reveals that organizations often interact for purposes outside the program's boundaries.

> Most of the folks working on the VSHP know each other from other things in the past and will continue to work together in other venues as well. These are overlapping organizations and groups that work together for different reasons. So [the VSHP] is one thing that pulls them together but it is not the only thing that pulls certain people to the tables.

For example, the Southern Tip Partnership comprises a subset of organizations involved in implementing the VSHP. This group focuses on acquiring and preserving land on the southern tip of the Eastern Shore. Other subgroups include the Birding and Wildlife Festival Committee, the Coastal Virginia Wildlife Observatory, the Eastern Shore Environmental Council, the Watershed Network, and the Avian Partnership. In this study, many of the same partners work together in different capacities and have done so for a number of years.

The findings suggest that high levels of trust play an important role in creating and sustaining collaborative interactions for two reasons. First, participants indicate that they feel comfortable with other members of the group because they know their partners will help them achieve the project's deliverables. Building this level of comfort is described during an interview:

> Trust is built through successful accomplishment of various projects that we work on and positive reinforcement. We build because we are at a point where we know the person will be there and they will follow through. Actions speak louder than words when working together.

As a result, trust allows organizations to rely on one another. This reliance is especially important in collaborative interactions because partners cannot individually achieve the same goals. An interview participant explains this reliance among researchers:

> In order to relate projects and do ecosystem wide research, you have to believe that other projects have value and that the research is trustworthy and that the people doing it know what they are doing. You can't go back and pick through the nitty gritty of their whole project, because you have to focus on yours. So that level of trust is vital to this system-wide approach.

These findings suggest that organizations involved in implementing the VSHP employ the "ethic of collaboration" discussed by Thomson and Perry (2006, p. 25). Interviewees indicate that they believe their partners are committed to the collective arrangement and will work in good faith with other organizations.

Second, trust plays an important role in creating and sustaining collaborative interactions because people become involved on a personal level. An interviewee discusses this personal involvement: "It is the fact that you see these folks all the time. The fact that it is a small landscape, very stable staff—people are here for a long time." As people learn more about one another, they know whom to call when they need help. Another interviewee attributes the development of personal relationships to the longevity of the network:

> After years of working with one another, you are no longer just working with an organizational face, but with a specific individual. The partnership evolves from an organizational relationship into a more personal relationship. You know who you need to call about certain things.

In addition to perceiving high levels of trust among partners within the network, interviewees suggest that they also feel high levels of trust with the VCZM Program. An interviewee explains this relationship: "The Coastal Zone Management Program has a long history on the Eastern Shore so this [i]s building upon or a reinvestment on past investments." Although sometimes outside the scope of the VSHP, the staff of the VCZM Program spends a great deal of time working with these same organizations. This suggests that high levels of trust between partnering organizations and the convening organization are needed to sustain collaborative interactions. As is suggested in the literature (see, for example, Bryson, Crosby, & Stone, 2006), high levels of trust are generated when organizations work together for long periods of time. For the VSHP, these organizations have worked together for 20 years to develop this high level of trust.

Organizational management is located at the cooperative end of the continuum in terms of risk taking. More than 85% of the elements identified within this variable suggest that there are low levels of risk associated with working together. An interviewee makes the following point pertaining to risk within the VSHP:

> There isn't really a risk involved. Each of us has the same objectives and goals. We are in the same arena. What the agency is really trying to accomplish works well for all of us...It behooves us to jump in and work together and help everyone. It is a benefit. It is another means to get where we need to be.

Another interviewee suggests: "[Risk] is minimal. Working together is helpful." High levels of risk based on dependency between organizations may be mitigated by long-standing relationships and a history that develops over a great deal of time.

The presence of these mitigating factors may explain this study's emphasis on low levels of risk. The relationship between risk and trust is described during an interview: "The risk is fairly low. And the trust is fairly high." Although the literature indicates that collaborative interactions involve high levels of risk (see, for example, Keast et al., 2004), this assertion is unfounded in this study. Long-standing relationships and high levels of trust are prerequisites for sustaining collaborative interactions. Despite the creation of dependencies between organizations, the likelihood of entering into a risky relationship seems far-fetched. An interviewee reveals how risk is minimized within the multiorganizational arrangement: "There are hard decisions to make, but the hesitancy is less when there are 20 other organizations standing with you saying that they agree and this needs to be done. And we show up for each other." These findings further support the removal of the risk-taking variable from the final version of the MIM.

Summary of Findings and Analysis

Based on the data collected through interviews and documents, administrators perceive interactions between organizations involved in the VSHP to be collaborative in nature. In this study, the interorganizational policy objective, interorganizational infrastructure, interorganizational procedures, and the organizational management constructs of the MIM are all found to operate at the collaborative end of the continuum of interaction. Outcomes from this research are used to analyze the implementation network, movement of multiorganizational relationships on the continuum of interaction, reconciliation of the top-down/bottom-up approaches to implementation, multiorganizational implementation strategies, and the application of collaborative interactions during policy implementation.

The Implementation Network

Outcomes from this study are used to analyze interactions between organizations implementing the Virginia Seaside Heritage Program. These interactions are explored by asking each interviewee the following question: "What organizations do you most closely work with to implement the Virginia Seaside Heritage Program?" When applicable, interview responses are corroborated with information in documents. An interview participant describes the multiorganizational arrangement as "a smorgasbord of organizations." Interactions between organizations in the VSHP are mapped in Figure 4.1.

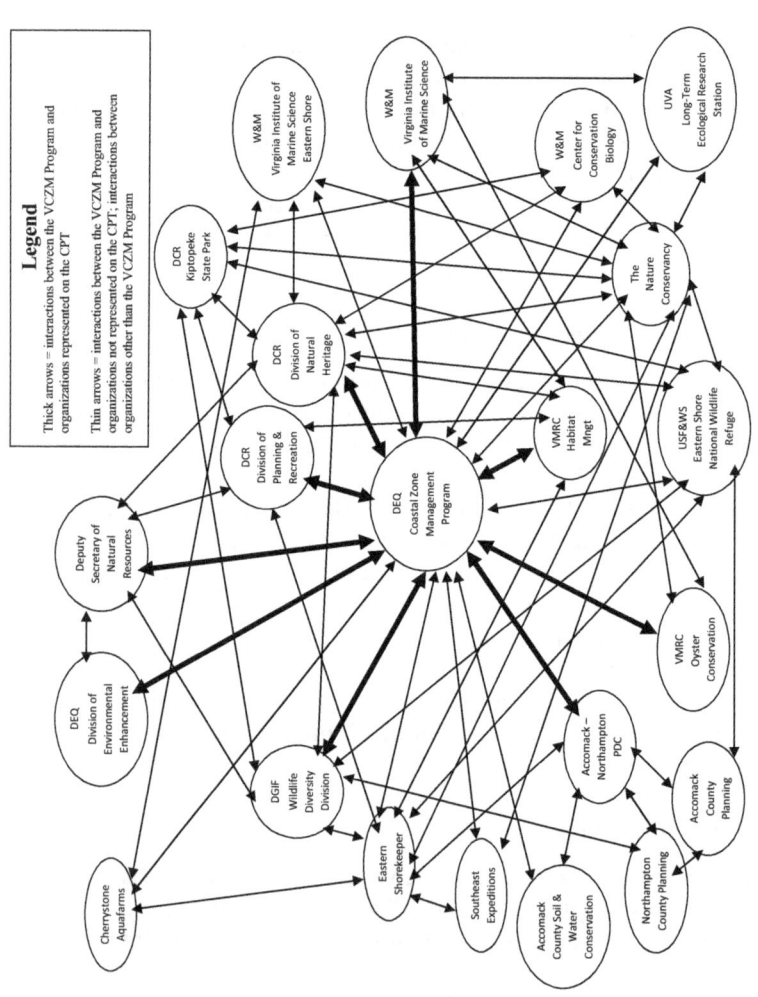

Figure 4.1 Virginia Seaside Heritage Program Implementation Network.

In the diagram of the VSHP implementation network, circles are used to represent participating organizations. In some instances, organizations are represented by more than one circle. This determination was based on an organization fulfilling one of the following factors: (1) multiple divisions within one organization are highly involved in the VSHP; (2) multiple divisions within one organization are represented separately on the Coastal Policy Team (CPT); or (3) multiple divisions within one organization are located in distinctly different geographic locations. Two types of arrows are used to convey interactions. Thick arrows represent interactions between the VCZM Program and organizations represented on the CPT. Thin arrows represent interactions between the VCZM Program and organizations not represented on the CPT. Thin arrows are also used to represent interactions between organizations that do not include the VCZM Program. Interviewees suggest that the strength of these interactions lie in their equality. When organizations interact to implement the VSHP administrators perceive these relationships to be of equal importance. These perceptions further emphasize that personnel involved in the VSHP see themselves as working among partners of equal status.

Mapping multiorganizational interactions is important because it helps identify organizations that play central roles during implementation of the VSHP. Although the diagram of the implementation network is not spatially oriented, organizations that play central roles are identified by looking at the number of connections they have with other organizations in the network. The organizations more central to the program's implementation are those with more arrows connecting them to other organizations.

The relative centrality of organizations within the implementation network is determined by comparing the number of organizational connections. Figure 4.2 is a spatial diagram that illustrates the results of this comparison. Organizations most central to the implementation network are the Department of Environmental Quality, Virginia Marine Resource Commission, Eastern Shorekeeper, U.S. Fish and Wildlife Service, Department of Conservation and Recreation, The Nature Conservancy, College of William and Mary, and Department of Game and Inland Fisheries. Federal/state agencies and nonprofit organizations comprise the majority of organizations within the core of the implementation network. Private organizations and local governments comprise the majority of organizations within the periphery.

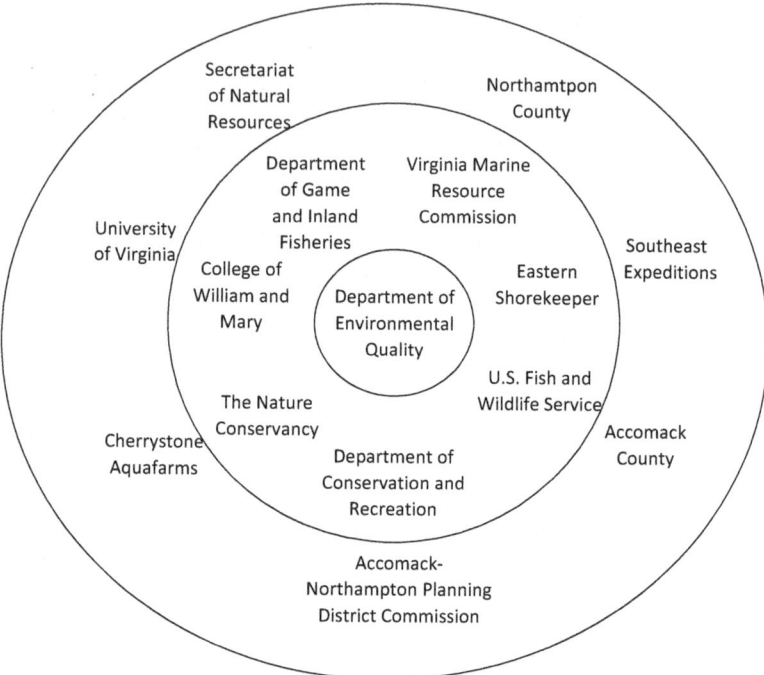

Figure 4.2 Relative Centrality of Organizations in the Implementation Network.

Perhaps the centrality of these organizations can be better understood when looking at the missions of the organizations involved. The centrality of federal/state agencies is not surprising given that they are required by executive order to participate in the VCZM Program in some capacity (Kaine, 2006). Evidence that this network operates beyond command-and-control authority comes from the centrality of nonprofit organizations. It is not a coincidence that the missions of The Nature Conservancy and Eastern Shorekeeper align holistically with the goals of the VSHP. Centralized roles within the implementation network are occupied by organizations whose missions most align with the goals of the program or policy. This finding suggests that collaborative interactions require mission alignment among organizations operating within the core of the implementation network. In addition, a common theme among interviewees is that nonprofit organizations play an important role in implementing the VSHP because they can operate in ways, and at speeds, that public organizations are unable to achieve. The presence of nonprofit organizations within the core of the network may be essential in developing and sustaining collaborative interactions.

Movement on the Continuum of Interaction

Data from this study are used to explore empirically the movement of multiorganizational relationships on the continuum of interaction. The following themes regarding the movement of these interactions emerge during analysis: (1) organizations appear to operate in some degree along all points of the continuum; (2) placement along the continuum may vary based on organizational function; and (3) relationships between organizations do not necessarily progress in a linear manner along the continuum.

First, data from this study does not support the assumption within the education literature that effective interactions between organizations are those in which the type of interaction is aligned throughout all aspects of the arrangement (see, for example, Thatcher, 2007). Although the aggregated presence of elements indicates that the implementation network operates at the collaborative end of the continuum for all four of the model's constructs, this pattern does not hold throughout the analysis of each interview and document. This suggests that organizations do not operate consistently within one type of interaction to the exclusion of all others. In fact, all three types of interactions are identified in 88% of interviews conducted for this study.

This also suggests that different organizations within the collective arrangement may use different types of interactions when working together. Of all the interactions between organizations, more than one quarter of them are between organizations that do not reciprocate the same type of interaction. Therefore, organizations may work together to implement policy but they may rely on different types of interactions to do so. For example, almost 90% of interviewees represent federal/state agencies and perceive collaborative interactions between organizations involved in implementing the VSHP. Data from this study indicates that 70% of federal/state agencies implementing the VSHP interact with local governments or nongovernmental organizations. However, only 25% of interviewees representing local governments and a little less than 50% of interviewees representing nongovernmental organizations perceive collaborative interactions between the organizations involved in implementing the VSHP. In this study, each local government and nongovernmental organization works with at least one federal/state agency during implementation.

Interactions that are not reciprocated may be explained when considering the centrality of organizations within the implementation network. Interactions between organizations with centralized roles seem to be reciprocated as collaborative in nature. In many instances, interactions are not reciprocated when organizations with centralized roles work with organizations that do not have centralized roles. In these instances, organizations central to the network seem to interact at a collaborative level while organizations less central to the network do not. This finding suggests that organizations central to the implementation network are more likely to perceive interactions to operate at the collaborative end of the continuum.

Second, data from this study suggests that the placement of an organization's interactions on the continuum may vary by function. For example, an interviewee explains that a particular organization's placement on the continuum of interaction is perceived to be in the coordinative area for administrative matters such as the exchange of money or the arrangement of meetings. However, this same organization is perceived to operate toward the collaborative end of the continuum when making decisions pertaining to the VSHP. While not always stated explicitly, this pattern emerges throughout other interviews. Participants directly involved with projects funded through grant money are more likely to mention elements associated with the coordinative area of the continuum. On the other hand, participants involved in the collective decision-making process of the CPT are more likely to mention elements associated with the collaborative end of the continuum. This finding suggests that personnel representing the same organization may perceive interactions differently depending on the functions they are involved in.

Third, data from this study suggests that relationships between organizations do not necessarily progress in a linear manner along the continuum. In 38% of the interviews, there is a nonlinear distribution of interactions along the continuum. These instances occur in the following two scenarios: (1) the number of elements associated with cooperative and collaborative interactions is higher than the number of elements associated with coordinative interactions; or (2) the number of elements associated with coordinative interactions is higher than the number of elements associated with cooperative or collaborative interactions. When aggregating the mention of elements in the interorganizational procedures and organizational management constructs of the MIM, the number of elements associated with cooperative and collaborative interactions are higher than the number of elements associated with coordinative interactions. In these constructs, more elements are linked to both ends of the continuum rather than creating a linear progression along the continuum.

The nonlinearity of these relationships may be explained when considering the maturity of relationships between organizations involved in the VSHP. It seems reasonable to suggest that a more linear progression along the continuum may occur in the beginning stages of the relationship as organizations become increasingly more interdependent. These findings suggest that once a relationship matures, multiorganizational relationships may move on the continuum to ensure operations align with the type of interaction needed to achieve the program's goals. Due to the amount of time and resources associated with collaboration, this type of interaction does not continue unless it is needed.

Reconciling Approaches to Implementation

This study supports the use of multiorganizational arrangements during policy implementation. Since administrators indicate that they continue to face resource shortages and complex problems, it is likely that multiorganizational arrangements will continue to be used in implementation. Therefore, the findings suggest that theorists may move beyond the top-down/bottom-up debate while acknowledging the strengths of each approach in two ways: (1) by focusing on the interactions within multiorganizational arrangements; and (2) by recognizing that multiorganizational implementation action occurs by linking organizations through middle-level personnel.

First, the two approaches historically used to approach implementation may be reconciled by focusing on multiorganizational interactions. The reason for this is that the variables within the MIM move along the continuum of interaction without being inhibited by the set of assumptions that guide either approach to implementation. Since the MIM considers both approaches to be equally relevant in exploring interactions between organizations, competition between the two approaches is unnecessary. For example, the impetus for collective action variable within the policy objective construct of the MIM has elements associated with the top-down and bottom-up approaches, but these elements are placed along different points of the continuum. At the cooperative end of the continuum, organizations initiate collective action because it helps build capacity within their world of work. Since these actions are typically initiated at the lower levels of the organization, this element may be associated with the bottom-up approach to implementation. As this variable moves along the continuum of interaction, more emphasis is placed on the top-down approach. Within the coordinative areas of the continuum, collective action is typically initiated through legislative mandates or grant contracts. An emphasis on policy mandates may be associated with the top-down approach to implementation. Despite this variable containing elements associated with competing implementation approaches, neither is assumed to be more important because each aligns with a different type of interaction.

This point is further conveyed as the variable moves toward the collaborative end of the continuum where collective action is initiated because no single organization can accomplish the task individually. At this point in the continuum, the top-down/bottom-up debate does not seem applicable. Regardless of the controls administered from the top or the dynamics occurring at the bottom of single organizations, interactions between organizations are needed to accomplish the task. Therefore, the top-down/bottom-up debate becomes less important as interactions move toward the collaborative end of the continuum.

Second, data from this study suggest that the top-down and bottom-up approaches to implementation may be reconciled by recognizing that multiorganizational implementation action occurs by horizontally linking organizations through middle-level personnel. Findings from this study support this assertion. Multiorganizational implementation within the VSHP emphasizes linking organizations through a horizontal structure in which all representatives are partners of equal status. Even though hierarchically structured government organizations represent 60% of the organizations participating in this implementation network, interactions between organizations involved in the VSHP are overwhelmingly collaborative. This finding suggests that participants involved in the VSHP transcend the hierarchical structures within their individual organizations and associate themselves with the horizontal structures of the collective arrangement. Two types of horizontal structures are used to establish linkages within the multiorganizational arrangement.

One type of horizontal structure within the VSHP comes from the CPT. Representatives from state and local governments participate in this group and make programmatic decisions to guide the program's overall direction. State agency representatives, who make up 83% of the CPT members involved in the VSHP, are resource administrators or managers selected to participate by the head of their agency.

Pooling field-level expertise through the creation of an executive steering committee creates a second type of horizontal structure in the VSHP. The people involved in this committee occupy operational positions and often spearhead projects associated with the VSHP. Their expertise is widely acknowledged by interviewees who hold positions on the CPT. These interviewees indicate that personnel making decisions at operational levels have 20–30 years of experience in studying the ecosystems on the Eastern Shore. The use of the CPT and executive steering committee facilitates the involvement of two levels of personnel from each state agency— resource administrators and operational project leaders. Representatives on the CPT typically supervise the project leaders on the executive steering committee. Both levels of personnel operate within the middle levels of their organizations. Neither the resource administrators nor the project leaders are at the very top or very bottom of their organizational structures. These findings suggest that multiorganizational implementation requires the development of horizontal connections within the collective arrangement in addition to the vertical connections within individual organizations. Furthermore, multiorganizational implementation may be optimal when horizontal connections are established at more than one level and when these connections are made between personnel occupying positions at the middle levels of organizations.

Multiorganizational implementation within the VSHP also emphasizes the involvement of middle-level personnel when it comes to making decisions. Based on the self-description from an interviewee, resource administrators on the CPT are considered "mid-level managers" when looking at their relative placement within individual agencies. Participants of the CPT see themselves as the right people to be involved in the program because they can quickly disseminate information throughout the organization when needed, and they can inform agency directors when problems arise at the operational level. Therefore, these personnel bridge gaps between the top and bottom of their respective organizations. They are perfectly situated to generate implementation action because they are high enough to commit resources to the collective effort and low enough to be aware of operational issues. Several interviews explain that the strength of the VCZM Program is based on the involvement of personnel who have the discretion to direct resources from their individual agencies toward projects aligned with the VSHP. Interviewees who serve as members of the CPT suggest that they make programmatic decisions and direct organizational resources based on feedback from personnel at the operational level.

Involvement of middle-level managers in the implementation network is also seen in the distribution of financial resources. The VCZM Program Manager, who holds a middle-level position in the Department of Environmental Quality, makes decisions regarding the distribution of grant funds and provides a significant source of leadership to the horizontal structures within the implementation network. This suggests that leadership within networks may come from personnel with positions in the middle of their organization's structure rather than from those at the top.

It appears that the nexus of the top-down/bottom-up approaches to implementation occurs because connections are made between middle-level personnel across different organizations who couple the two approaches. They convey the policy mandate to those at the operational level, they ensure top management is aware of operational problems, and they make resource decisions that comply with the policy mandate while supporting operational needs. The involvement of middle-level personnel is described by an interviewee:

> What ends up functionally occurring is that in the middle is where everything happens. That is true for people. The people at the top get called when something goes wrong or when there is political pressure. The people at the bottom don't necessarily have all the connections yet. So they might be good working one on one. But they aren't the people who ensure that the funding stays in place. [This occurs through] the people in the middle.

Another interview participant associates the program's success with the involvement of resource administrators: "...one of the more effective aspects of [the program is] that the people directly responsible for management of resources pretty much get to decide where we get to provide our focus without a whole lot of political oversight." Agreement is expressed by another interviewee:

> The Coastal Policy Team is comprised of the right level of people. We are at the right level in the organization where we can quickly disseminate top-down if a particular problem or program policy arises. We are able to quickly get that down into the organization because it is one of our primary responsibilities.

But it works just as well the other way. If a problem or need is identified at the program level or a solution comes to light, we are just as easy to talk to the chief deputy or the director at any time.

These findings suggest that utilizing multiorganizational implementation may be most conducive when the following conditions are met: (1) middle-level managers are involved; (2) these managers have the discretion to allocate organizational resources to collective efforts; and (3) these managers have the expertise and time to understand issues at the operational level.

Multiorganizational Implementation Strategies

The findings from this study suggest that theoretical understanding of policy implementation in multiorganizational arrangements may be improved through the exploration of a range of interactions. Through the use of the MIM, this study is the first to link cooperation, coordination, and collaboration collectively to multiorganizational implementation. By looking at the entire continuum of interactions, researchers are better able to identify the conditions under which it is appropriate to use a particular type of interaction as an implementation strategy. These conditions may be identified by looking at the elements most emphasized within each type of interaction. The presence of certain activities, such as those described in Table 4.3, provide some of the conditions in which it may be appropriate to utilize a particular type of interaction. For example, if dialogue can be maintained through informal relationships,

Table 4.3 Implementation Strategies: Elements Most Emphasized in Each Type of Interaction

Cooperation	Coordination	Collaboration
Dialogue maintained through informal relationships	Legislative mandate or grant contracts enhance cohesion or minimize duplication	History of supportive behavior or long-standing relationships
Informally work together to achieve individual goals	Resources may be provided through mandate or grant arrangements	A lead agency or convener brings relevant stakeholders together

(Continued)

Cooperation	Coordination	Collaboration
Interest of individual organization paramount	Linkages are mobilized because compatible mission areas mutually increase abilities to achieve individual goals	Complementary interests in attaining mutual goals
Independent; possible to accomplish the task individually	Mechanisms, such as contractual or nonfinanical agreements, formalize relationships	Participative decision making through consensus and compromise
Work completed as part of regular job responsibilities	Statutes or grant contracts provide funding	Understanding is enhanced by a willingness to share information about organizations, which may include what can/cannot be offered to the collective group

then cooperative interactions may be an appropriate implementation strategy to use. On the other hand, if legislative mandates or grant contracts are needed to enhance cohesion then coordinative interactions may be an appropriate implementation strategy to use. Furthermore, if there is a history of supportive behavior or long-standing relationships, then collaborative interactions may be an appropriate implementation strategy to use.

Collaborative Interactions during Implementation

Findings from this study reiterate that collaboration is not appropriate for use in all situations. Participants stress that interactions between partners involved in implementing the VSHP require great amounts of time and resources to sustain. The conditions identified within the literature as being conducive for collaboration are supported by this research. In this study, 44% of interviewees indicate that there is a sense of environmental crisis; 50% of interviewees recognize that their organizational responsibilities toward coastal management on the Eastern Shore interconnect with those of other organizations; and 76% of interviewees indicate that organizations have mutual interests. Furthermore, over 80% of the interview participants mention the importance of trust between partners.

This suggests that the development of trust between organizations is another condition that may determine the extent to which a situation lends itself to collaborative interactions. Relationships between partners involved in implementing the VSHP are enhanced by 20 years of working together. Therefore, collaborative interactions should not be expected to develop quickly or easily because they take great effort from all organizations involved. Findings from this study confirm that certain circumstances are more conducive to collaborative interactions.

Analysis of Initiation of Interactions

Variables within the MIM are analyzed to address the following research question: How are multiorganizational interactions initiated? This research question guides inquiry into whether interactions are initiated formally through legislative mandate or agency rulemaking, informally through street-level experience or common interests, or a combination of both. As textual data are collected from interviews and documents, the researcher uses content analysis to organize the data into categories of the pre-structured coding scheme. Elements in the impetus for collective action variable within the interorganizational policy objective construct and the formality of the agreement variable within the interorganizational infrastructure construct are explored to respond to this research question. The following two questions are asked during interviews: (1) What brought the organizations together to implement the VSHP? (2) How are the roles and responsibilities for each participating organization determined?

Analysis of Formally Initiated Interactions

Textual data from interviews and documents suggest that organizational interactions are formally initiated and sustained within the VSHP. Formal interactions within the VSHP are most prevalent when participants speak of grant contracts. As money changes hands, clear roles and responsibilities are delineated for the organization awarded the grant. Due to an emphasis on creating horizontal linkages and involving middle-level personnel, the accountability mechanisms typically generated by bureaucratic organizations do not appear in multiorganizational arrangements. Therefore, these formalized interactions fill an important gap in creating accountability mechanisms for the grantee.

The prevalence of formal interactions is consistent with much of the empirical inquiry within the multiorganizational implementation literature (see, for example, Hall & O'Toole, 2004; O'Toole, 1995). However, much of this literature focuses on interactions initiated by policy mandate, agency rulemaking, or organizational procedures. More specifically, the literature focuses on the extent to which policies identify organizational partners, policy characteristics that induce or constrain interdependence, or the structures used in multiorganizational implementation (see, for example, Hall & O'Toole, 2004; May, 1995; Raelin, 1982). Some organizational documents pertaining to the VSHP identify multiorganizational partners and the presence of the CPT, but these documents do not make any specifications regarding the ways in which organizations should work together. In this study, interviewees do not perceive formal interactions to be initiated through the methods specified in the literature. For example, not one interview participant cites a policy mandate as the impetus for organizations working together. Instead, 64% of interview transcriptions and documents indicate that interactions are initiated through grant contracts. An interviewee describes the relationship between funding and multiorganizational interactions: "Partnering happens as a result of funding. It gives us something to work with." Another participant expresses agreement: "The money initially br[ings] all of us together." Findings from this study suggest that formal interactions are initiated by grant contracts rather than policy mandates, agency rulemaking, or organizational procedures.

Another 43% of the interview transcriptions and documents indicate that roles are formalized through grant contracts. The formalization of roles is revealed during an interview: "Our roles for a particular project are defined by the terms of the grant. As a grantee, our role is clear. We have a grant document that we agree to and we have to live up to those terms." Discussion during another interview reveals: "There are certain rules stipulated within the grant contracts." An example of this formalization of roles can be seen in the relationship between the VCZM Program and the PDCs. Each year, the VCZM Program provides a grant fund to each of the PDCs. Along with this grant money, the staff of the VCZM Program identifies a minimum standard for what each PDC has to do in return for these funds. For example, they are required to conduct quarterly meetings and training sessions with local government administrators. According to an interviewee, these requirements "provide the conduit for flow of information from the state through the PDC to the localities and just as importantly from the localities back up to the state." These findings align with the research conducted by Van de Ven and Walker (1984) and suggest that interdependencies generated through financial resources rely on a formal approach to interaction.

Analysis of Informally Initiated Interactions

Textual data from interviews and documents suggest that organizational interactions are also informally initiated and sustained within the VSHP. Informal interactions within the VSHP are most prevalent when participants speak of working with peers to enhance their abilities to achieve organizational goals. Informal interactions occur among the organizations involved in the VSHP, but they seem to be less prevalent than the formal interactions induced by grant contracts. Consistent with the research conducted by Van de Ven and Walker (1984), ad hoc relationships are identified in this study when partners align resources. An interviewee explains the presence of informal relationships: "The Coastal Zone Management Program officially pull[s] us all together. But we all started talking together informally long before we came up with the nuts and bolts of it." A common theme among discussions during interviews is that operational personnel become involved in the VSHP for two reasons: (1) they are asked by another partner; or (2) they think it will be beneficial to their world of work.

First, field-level personnel often work on projects associated with the VSHP because other partners pull them in. An interviewee describes these connections within the multiorganizational arrangement: "Organizations are pulled in as necessary by working through the Coastal Policy Team. They are contacted when needed." Another interviewee explains, "A partner recently called me and asked if we wanted to be involved in a particular project. I called the Coastal Zone Management Program and asked if they wanted to jump in on this as well." Discussion during another interview suggests that these contacts are possible because of long-standing relationships and information sharing among partners. "Having those long-standing relationships really helps in terms of pulling the partners together. The partners themselves pull in extra people when they need to." It is important that personnel involved in the VSHP understand the mission and interests of other organizations because a high level of understanding helps them know who to call.

Second, operational personnel become involved in the collective arrangement because it is beneficial to their world of work. This theme is prevalent during interviews. An interview participant indicates that organizations become involved in the VSHP because it creates opportunities to utilize expertise and research in new ways:

> Our involvement came from us. We ha[ve] been working on Phragmites since the mid-1990s. For us, it seem[s] that it [i]s an opportunity to work on Phragmites in a place that is high priority for us and to go at it in a manner that we ha[ve] not had an opportunity before.

Summary of Findings and Analysis

The data suggests that interactions between organizations involved in implementing the VSHP are initiated formally and informally. The findings from this research are important for two reasons: (1) multiorganizational policy implementation occurs in part through informal relationships; and (2) a majority of interviews perceive formalized interactions to be initiated through grant contracts. First, this study is the first to acknowledge that multiorganizational policy implementation occurs in part through informal relationships. Although informally initiated interactions are less prevalent during implementation of the VSHP, their presence is important because it suggests that interactions are initiated in ways other than the current body of literature explores. In addition, the presence of informally initiated interactions during multiorganizational implementation suggests that a top-down approach does not fully capture relationships that occur outside the boundaries of operational authority.

Second, a majority of interviews perceive formalized interactions to be initiated through grant contracts. The literature emphasizes formalized interactions deliberately configured through policy mandates, agency rulemaking, or organizational procedures. Although interactions are formally initiated, the findings from this study suggest that the literature's approach to formalized interactions may need to be reconsidered. Interactions within the VSHP are not necessarily the result of policy mandates, agency rulemaking, or organizational procedures. Despite the presence of a policy mandate encouraging organizations to work together, not one interviewee cites this mandate as the impetus for collective action.

Instead, 64% of interviews and documents indicate that interactions are initiated through grant contracts.

References

Agranoff, R. (2006). Inside collaborative networks: Ten lessons for public managers. *Public Administration Review, 66.* 56–65.

Bryson, J., Crosby, B., & Stone, M. (2006). The design and implementation of cross-sector collaborations: Propositions from the literature. *Public Administration Review, 66.* 44–55.

Hall, T., & O'Toole, L. (2004). Shaping formal networks through the regulatory process. *Administration and Society, 36*(2). 186–207.

Imperial, M. (2001). Collaboration as an implementation strategy: An assessment of six watershed management programs. *Dissertations & Thesis Full Text, 62*(02), 767. (UMI No. 3005481)

Imperial, M. (2005). Using collaboration as a governance strategy: Lessons from six watershed management programs. *Administration & Society, 37*(3). 281–320.

Jennings, E. (1994). Building bridges in the intergovernmental arena: Coordinating employment and training programs in the American states. *Public Administration Review, 54*(1). 52–60.

Jennings, E., & Krane, D. (1994). Coordination and welfare reform: The quest for the philosopher's stone. *Public Administration Review, 54*(4). 341–348.

Kaine, T. (2006). *Executive order number twenty-one.* Retrieved February 4, 2008, from http://www.deq.state.va.us/coastal/exorder.html

Keast, R., Brown, K., & Mandell, M. (2007). Getting the right mix: Unpacking integration meanings and strategies. *International Public Management Journal, 10*(1). 9–33.

Keast, R., Mandell, M., Brown, K., & Woolcock, G. (2004). Network structures: Working differently and changing expectations. *Public Administration Review, 64*(3). 363–371.

Mandell, M. (1994). Managing interdependencies through program structures: A revised paradigm. *American Review of Public Administration, 24*(1). 99–121.

Mandell, M., & Steelman, T. (2003). Understanding what can be accomplished through interorganizational innovations: The importance of typologies, context, and management strategies. *Public Management Review, 5*(2). 197–224.

Mattessich, P., Murray-Close, M., & Monsey, B. (2001). *Collaboration: What makes it work?* Saint Paul, MN: Amherst H. Wilder Foundation.

May, P. (1995). Can cooperation be mandated? Implementing intergovernmental environmental management in New South Wales and New Zealand. *Publius, 25*(1). 89–113.

McNamara, M. W., Leavitt, W., & Morris, J. (2008). *Multiple-sector part-nerships and the engagement of citizens in social marketing campaigns.* Paper for Annual Conference of the American Society for Public Administration, Dallas, TX.

O'Toole, L. (1995). Rational choice and policy implementation: Implications for interorganizational network management. *American Review of Public Administration*, 25(1). 43–57.

Raelin, J. A. (1982). A policy output model of interorganizational relations. *Organization Studies*, 3(3). 2243–267.

Thatcher, C. (2007). A study of interorganizational arrangement among three regional campuses of a large land-grant university. *Dissertations & Thesis Full Text*, 68(03). (UMI No. 3255178)

Thomson, A., & Perry, J. (2006). Collaboration processes: Inside the black box. *Public Administration Review*, 55. 20–32.

Van de Ven, A., & Walker, G. (1984). The dynamics of interorganizational coordination. *Administration Science Quarterly*, 29(4). 598–621.

Wood, D., & Gray, B. (1991). Toward a comprehensive theory of collaboration. *Journal of Applied Behavioral Science*, 27(2). 139–162.

5 Implications for Theory and Practice

An enduring question in the field of interorganizational study concerns the ways different organizations interact with one another. The patterns of interaction, and the bases by which they occur, become the focus of scholarly attention. Moreover, when those organizations are located in different sectors of society, understanding the patterns of interaction becomes critical. As stated at the outset of this book, different terms are often applied to interorganizational interactions, but there has been little attempt to delineate between these definitions. The result is conceptual ambiguity, confusion, and an inability to examine these interactions in a systematic manner. This study examines those interactions in an attempt to better delineate, and thus distinguish, between different types of interactions. In short, how do we differentiate cooperation from coordination from collaboration? If these terms are used interchangeably, then the terms themselves become useless as descriptors of interorganizational behavior. On the other hand, if these terms have meaning apart from the others, then we must be able to understand how they are different. Our work in this book attempts to unravel the differences between these terms and bring conceptual clarity to each term.

Summary of Research

The purpose of this research is to explore interactions between organizations when working together to implement policy. The Multiorganizational Interaction Model (MIM) is presented as the theoretical basis for exploring the use of cooperation, coordination, and collaboration between government and nongovernmental organizations during implementation of the VSHP. This study is important because it marks the first time that a model linking the policy implementation and interorganizational theory literatures is used to explore empirically different types of interactions.

The Multiorganizational Interaction Model

This study is guided by a research question that focuses on the helpfulness of the Multiorganizational Interaction Model (MIM) in explaining interactions in a policy implementation setting. A response to this question is formulated by organizing textual data into the categories of a predetermined coding scheme aligned with the model's operationalizations. Elements aligned with each of the following four constructs of the model are mentioned in interviews and documents: interorganizational policy objective, interorganizational infrastructure, interorganizational procedures, and organizational management.

Alignment between the data's empirical patterns and the model's theoretical patterns suggest that the revised model is helpful in explaining interactions between organizations during multiorganizational implementation. We have established authenticity for the MIM in two ways. First, more than 92% of the elements introduced in the model are identified in this study. These elements operationalize the variables within the MIM and align with the policy implementation and interorganizational theory literatures. Suggestions are made in Chapter 3 to address eight elements from the model that are not identified in interviews or documents. One half of these elements are removed from the final version of the model while the other half of the elements remain for further research. Second, authenticity of the model is further established as all patterns identified in interviews and documents fit into one of the categories established in the predetermined coding scheme. Although an opportunity to gather competing evidence is given, interviewees do not mention factors that are not already captured within the MIM.

Interactions during Multiorganizational Implementation

This study explores a second research question pertaining to how administrators perceive the use of cooperation, coordination, and collaboration when working in a multiorganizational arrangement to implement policy. The perceived use of these interaction terms in this study are determined by organizing the data into the categories of the predetermined coding scheme aligned with the operationalizations of the MIM. Elements align with each of the following three types of interactions: cooperation, coordination, and collaboration.

Interactions between organizations involved in implementing the Virginia Seaside Heritage Program are perceived to be highly collaborative in both interviews and documents. In this study, the interorganizational policy objective, interorganizational infrastructure, interorganizational procedures, and organizational management constructs of the MIM are all found to operate at the collaborative end of the continuum of interaction. Despite a policy mandate that requires organizations to work together on coastal zone issues and the presence of documents intending to formalize multiorganizational relationships, elements associated with collaborative interactions are emphasized twice as often as elements associated with cooperation or coordination. This study supports the assertion that theorists may reconcile the top-down/bottom-up approaches by focusing on interactions within multiorganizational arrangements. Progress toward a fourth generation of implementation research is made by recognizing that implementation action occurs by linking organizations through middle-level personnel. Administrators implementing the VSHP perceive interactions between organizations to operate beyond formalized mechanisms at an overwhelmingly collaborative level. This finding is especially interesting because government organizations represent a majority of the organizations involved in this study. It suggests that multiorganizational implementation requires the development of horizontal connections between organizations in addition to vertical connections within organizations. Government employees transcend highly centralized and hierarchical structures to create and sustain horizontal linkages between organizations. In this study, linkages develop across organizational boundaries through middle-level personnel. Implementation action occurs because these personnel are perfectly situated to commit resources to the collective effort while being aware of operational issues. Multiorganizational implementation is optimal when organizations are horizontally linked through middle-level personnel at multiple levels.

In addition, elements associated with each of the three types of interactions are identified in this study. Although interactions between organizations implementing the VSHP are collaborative, elements associated with cooperative and coordinative interactions are also identified. Two themes regarding the movement of interactions between organizations on the continuum emerge. First, organizations appear to operate in some degree along all points of the continuum. Relationships between organizations do not operate consistently within one type of interaction to the exclusion of all others; all three types of interactions are identified in a majority of the interviews. As organizations work together to implement policy, they rely on different types of interactions. Second, the perceived placement of multiorganizational interactions along the continuum may vary based on organizational function. Participants directly involved with projects funded through grant money are more likely to mention elements associated with the coordinative area of the continuum while participants involved in the collective decision-making process of the Coastal Policy Team are more likely to mention elements associated with collaborative interactions.

The findings from this study reiterate that collaboration is not appropriate for use in all situations. Participants involved in implementing the VSHP indicate that developing and sustaining the collaborative arrangement take great amounts of time and resources. This study confirms that certain factors lend themselves to collaborative interactions. The presence of a sense of crisis, interconnected responsibilities, mutual interests, and trust may help situations become more conducive for collaborative interactions.

The Initiation of Interactions

Our final research question examines how multiorganizational interactions are initiated. Textual data are organized into the categories of a predetermined coding scheme aligned with the operationalizations of the MIM. Elements within the impetus for collective action variable and the formality of the agreement variable are used to explore this question.

Interactions between organizations implementing the VSHP are initiated formally and informally. Although much of the literature suggests that formalized interactions are initiated by policy mandates, agency rulemaking, or organizational procedures, formally initiated interactions in the VSHP are most prevalent in grant contracts. It is through these formalized interactions that accountability mechanisms are created within the multiorganizational arrangement. Since the accountability mechanisms typically found in bureaucratic organizations do not appear in the implementation network, these formally initiated interactions fill an important gap. This finding is important because it suggests that interactions during multiorganizational implementation are formally initiated in ways other than the current body of literature suggests.

In addition, formally initiated interactions are important in multi-organizational arrangements because they generate accountability mechanisms for the distribution of financial resources.

Multiorganizational interactions are also informally initiated within this study. These interactions are most prevalent when participants speak of working with their peers to achieve organizational goals or align resources. Understanding the missions and interests of partnering organizations helps personnel informally develop and sustain relationships. This finding is important because it suggests that multiorganizational implementation occurs, in part, through informally initiated interactions.

Contributions to Interorganizational Theory

The literature often describes interactions within multiorganizational arrangements as cooperative, coordinative, or collaborative. These descriptions are problematic because researchers use them interchangeably, without regard for potential distinctions of meaning between the terms. Findings from this study support the assertion that each type of interaction is independent and different from the others. This study makes two contributions to interorganizational theory. First, a major contribution of this study to the interorganizational theory literature is the development of the MIM. Ambiguities within a model previously used to explore interactions between organizations limited its applicability. The MIM clearly distinguishes between the three types of interaction: cooperation, coordination, and collaboration. Different elements are associated with each interaction based on the policy implementation and interorganizational theory literatures. Distinctions between different types of interactions are important because theorists may now subject this model to empirical testing in settings outside of policy implementation. Varied application will benefit the model and enhance its transferability.

In turn, improvements to the model may also generate theoretical consistency and improve communication within the interorganizational theory literature. Up to this point, theorists have not embraced a model that allows them to acknowledge different types of interactions between organizations collectively. Instead, they tend to focus on multiorganizational interactions in a singular way and assume that a specific type of interaction occurs. The MIM creates a foundation for comparing interaction terms empirically. Since a specific type of interaction will not be effective in all settings (Keast, Brown, & Mandell, 2007; Thomson & Perry, 2006), it is especially important to understand distinctions between them. The constructs and variables within the theoretical model provide a needed structure for deciphering between different interaction terms.

Second, a contribution of this study to the interorganizational theory literature involves clarifying the movement of interactions between organizations on the continuum. While some researchers describe cooperation, coordination, and collaboration as falling along a continuum of increased interaction (see, for example, Mattessich, Murray-Close, & Monsey, 2001; Thomson & Perry, 2006), this is the first time that the movements of interactions along the continuum are explored empirically. Relationships between organizations do not necessarily progress in a linear manner along the continuum. This may be a result of the maturity of the relationships between the organizations involved in the VSHP. A more linear progression along the continuum may occur in the beginning stages of relationships due to increased levels of interdependence. As relationships mature, they may move on the continuum to ensure operations align with the type of interaction necessary to achieve the program's goals. Nonlinear movements along the continuum of interaction support assertions within the literature that one type of interaction is not inherently better than the others. Findings from this study suggest that nonlinear movements give organizations flexibility to adjust to contextual conditions by moving to a different area along the continuum.

Contributions to Policy Implementation Inquiry

Much of the previous policy implementation research focuses on differences between the top-down and bottom-up approaches. This line of inquiry is problematic for theoretical advancement because an emphasis on synthesizing the top-down and bottom-up approaches fails to account for the multiorganizational arrangements frequently involved in policy implementation. Textual data gathered from interviews and documents supports the use and emphasizes the importance of multiorganizational arrangements during policy implementation. This study stands apart from previous research because it examines interactions during multiorganizational implementation empirically. By broadening the scope of current inquiry, this research contributes to theoretical inquiry in the policy implementation literature in four ways.

First, the findings from this study may help implementation theorists move beyond the top-down/bottom-up debate. The literature may be guided toward a fourth generation of implementation research by focusing on the interactions within multiorganizational arrangements. The MIM embraces elements associated with competing implementation approaches, but neither approach is assumed to be more important because elements align with different types of interactions. The top-down/bottom-up debate becomes less important as interactions move toward the collaborative end of the continuum. As multiorganizational arrangements become increasingly prevalent, implementation success may have more to do with how well organizations work together rather than the specificity of policy characteristics or the acknowledgment of environmental conditions at the local level. It is important to understand the linkages within a multiorganizational arrangement, and the MIM provides a theoretical lens to explore these linkages.

Second, theoretical understanding of policy implementation in multiorganizational arrangements may be improved through the exploration of a range of interactions. Through the use of the MIM, this study is the first to link cooperation, coordination, and collaboration collectively to multiorganizational implementation. By looking at the entire continuum of interactions, researchers are better able to identify the conditions under which it is appropriate to use a particular type of interaction as an implementation strategy.

Third, the literature may be guided toward a fourth generation of implementation research through the realization that implementation may occur by connecting middle-level personnel across different organizations. Although the public administration literature typically associates government organizations with highly centralized and hierarchical structures, the findings suggest that structures within individual organizations are far less important to the administrators implementing the VSHP than the horizontal structures linking organizations. It appears that the nexus of the top-down/bottom-up approaches to implementation occurs because middle-level personnel across different organizations couple the two approaches.

Fourth, theoretical understanding of policy implementation in multiorganizational arrangements may be expanded through the exploration of the ways interactions are initiated. Interactions between organizations are initiated in ways other than the literature suggests. The literature emphasizes formal interactions deliberately configured through policy mandates, agency rulemaking, or organizational procedures. However, findings from this study suggest that formalized interactions within the VSHP are initiated through grant contracts. The literature's focus on the origin of formalized interactions should be reconsidered.

This study is the first to acknowledge that multiorganizational policy implementation occurs in part through informal relationships. Although informally initiated interactions are less prevalent than formal interactions, their presence in this study suggests that multiorganizational implementation occurs outside the boundaries of operational authority. This line of inquiry is not considered in the current literature. Therefore, the presence of interactions informally initiated within this study's implementation network suggests that multiorganizational arrangements should not be treated as a mere extension of hierarchical organizations that abide by specifications in policy mandates. By focusing on formally initiated interactions as the sole source for action, researchers miss a piece of the larger picture. The different ways in which arrangements are initiated should continue to be examined empirically. This gap in current inquiry highlights a need for fourth generation implementation research to move beyond the top-down/bottom-up debate.

But, What about Practitioners?

Of course, theory development can be useful to scholars of interorganizational theory, in that it spurs the development of additional theory. However, we must not lose sight of the fact that these interactions occur in real-world settings. How does this research contribute to the practice of interorganizational implementation?

First, those engaged in interorganizational interactions can apply the lessons of this research to better structure their interactions with their partner organizations. This begins with the obvious realization that not all organizations are the same. From this, to achieve an understanding of how organizations differ can better align our expectations for the partner organizations. What sort of authorities do the organizations possess? What is the expected time horizon of the interaction? Is participation on the part of an organization voluntary, or has it been mandated by some external authority? What is the nature of the task to be accomplished? By thinking about these kinds of questions, drawn from the elements of the MIM, practitioners can better tailor their actions to a pattern of interaction that offers the best chance of success given the choice of partners.

Second, the MIM can be employed to gain a greater understanding of both the possibilities, and the limitations, of partner organizations. By matching our expectations for the interaction to the possibilities offered by the combination of partners and opportunities, we can work to build the kind of interaction that will not only provide for a satisfying and productive partnership but also for a successful attainment of goals. In short, the odds of success are greatly increased if the interaction pattern employed fits the participants and the collective goals.

Finally, practitioners can use the MIM to educate partners, potential or current, to the issues of successful interaction. Partners who have a clear sense of the possible, along with a firm understanding of both the strengths and limitations of a particular form of interaction, will ultimately make better partners. Concomitantly, by carefully considering the nature of our partners and the kinds of interactions that make sense, we avoid wasted time and resources by chasing interactions in which the partners have very different ideas about the nature of the interaction at hand. For example, both partners may refer to an interaction as "collaboration," certainly a desirable "buzzword" among organizations, but if one organization is pursuing what is effectively cooperation while the other is pursuing collaboration, the chances for failure increase substantially. If the meanings are shared, the chances for *success* increase substantially.

Future Research

Due to the use of a single case study design, findings from this research may be specific to interactions between organizations implementing the VSHP. While the model is found to be useful in explaining multiorganizational interactions in this setting, future research should focus on using the MIM to explore interactions between organizations in other settings. The theoretical model's confirmability may be enhanced through additional research. For example, relationships between constructs should be further explored. While this research suggests that relationships between variables within the organizational management construct are present, relationships between variables from different constructs are not prevalent.

The findings from this study suggest that the literature should continue to refocus its energies away from the top-down/bottom-up debate toward more fruitful lines of inquiry like multiorganizational interactions. As multiorganizational arrangements become increasingly prevalent, future research should continue to explore multiorganizational implementation. In these situations, implementation success may have more to do with how well organizations work together rather than the specificity of policy characteristics or the acknowledgment of environmental conditions at the local level. Therefore, continued exploration of linkages within multiorganizational arrangements is necessary.

Informal relationships play a significant role in developing and sustaining relationships between organizations implementing the VSHP, and this is an area deserving of further inquiry. More specifically, informally initiated interactions can be explored through cooperative relationships. While almost 20% of the elements mentioned throughout interviews and documents are associated with cooperative interaction, empirical research for this type of interaction is visibly absent from the interorganizational theory literature. The potential for interactions to be informally initiated should be further explored through research focused on cooperative interactions. Theoretical understanding of policy implementation in multiorganizational arrangements may be improved through the exploration of the ways interactions are initiated.

Findings from this study also indicate that coordinative interactions are most often associated with grant processes in which some organizational roles and responsibilities are formalized. As organizations work across organizational boundaries, the use of hierarchical controls becomes less relevant and formalized structures are often associated with the distribution of money. The applicability of coordinative interactions in multiorganizational arrangements may need to be reconsidered in order to explore the context of grant processes and their influence on formalized interactions.

In addition, previous research suggests that policy mandates infrequently provide monetary support to the organizations designated to implement them (see, for example, Montjoy & O'Toole, 1979; O'Toole & Montjoy, 1984). The Coastal Zone Management Act is considered an "unfunded mandate." While a stable funding stream is created through the distribution of grant funds, the availability of funds is less likely in other situations. Many policy mandates do not have a state agency distributing one half of a million dollars to entice organizations to work together. The criticality of a stable funding stream in this study suggests that researchers should consider ways in which implementation networks can access funding streams despite the regularity of unfunded mandates.

Furthermore, data from this study suggests that the convening organization plays a pivotal role in bringing organizations together and providing an organizational structure to carry out the program's operations. Although this role is acknowledged in the literature (see, for example, McNamara, Leavitt, & Morris, 2008), specific guidance for convening organizations is lacking. Additional research should further explore the roles of convening organizations.

References

Keast, R., Brown, K., & Mandell, M. (2007). Getting the right mix: Unpacking integration meanings and strategies. *International Public Management Journal*, 10(1). 9–33.

Mattessich, P., Murray-Close, M., & Monsey, B. (2001). *Collaboration: What makes it work?* Saint Paul, MN: Amherst H. Wilder Foundation.

McNamara, M. W., Leavitt, W., & Morris, J. (2008). *Multiple-sector partnerships and the engagement of citizens in social marketing campaigns.* Paper for Annual Conference of the American Society for Public Administration, Dallas, TX.

Montjoy, R., & O'Toole, L. (1979). Toward a theory of policy implementation. *Public Administration Review*, 39(5). 465–476.

O'Toole, L., & Montjoy, R. (1984). Interorganizational policy implementation: A theoretical perspective. *Public Administration Review*, 44(6). 491–503.

Thomson, A., & Perry, J. (2006). Collaboration processes: Inside the black box. *Public Administration Review*, 55. 20–32.

Index

3 20